String Theory and the Scientific Method

String theory has played a highly influential role in theoretical physics for nearly three decades and has substantially altered our view of the elementary building principles of the universe. However, the theory remains empirically unconfirmed, and is expected to remain so for the foreseeable future. So why do string theorists have such a strong belief in their theory?

This book explores this question, offering a novel insight into the nature of theory assessment itself. Dawid approaches the topic from a unique position, having extensive experience in both philosophy and high energy physics. He argues that string theory is just the most conspicuous example of a number of theories in high energy physics where non-empirical theory assessment has an important part to play. Aimed at physicists and philosophers of science, the book does not use mathematical formalism and explains most technical terms.

RICHARD DAWID is a Philosopher of Science at the University of Vienna, and has a PhD in theoretical physics. His main research interests are philosophical questions related to contemporary high energy physics, as well as general issues in the philosophy of science.

String Theory and the Scientific Method

RICHARD DAWID
University of Vienna

CAMBRIDGE
UNIVERSITY PRESS

University Printing House, Cambridge CB2 8BS, United Kingdom

One Liberty Plaza, 20th Floor, New York, NY 10006, USA

477 Williamstown Road, Port Melbourne, VIC 3207, Australia

314-321, 3rd Floor, Plot 3, Splendor Forum, Jasola District Centre, New Delhi - 110025, India

79 Anson Road, #06-04/06, Singapore 079906

Cambridge University Press is part of the University of Cambridge.

It furthers the University's mission by disseminating knowledge in the pursuit of education, learning and research at the highest international levels of excellence.

www.cambridge.org
Information on this title: www.cambridge.org/9781107449619

© R. Dawid 2013

First published 2013
Paperback edition first published 2015

A catalogue record for this publication is available from the British Library

Library of Congress Cataloging in Publication data
Dawid, Richard, 1966–
String theory and the scientific method / Richard Dawid, University of Vienna.
pages   cm
Includes bibliographical references and index.
ISBN 978-1-107-02971-2 (hardback)
1. String models.   2. Science – Methodology.   I. Title.
QC794.6.S85D39   2013
539.7′258–dc23
2012044103

ISBN 978-1-107-02971-2   Hardback
ISBN 978-1-107-44961-9   Paperback

To Walter (1937–2008)

Contents

Preface

This book has been on my mind ever since I left physics and turned to philosophy in the year 2000. A core motivation for making that step at the time was my feeling that something philosophically interesting was going on in fundamental physics but remained largely unappreciated by the world outside the departments of theoretical physics – and underappreciated even within. Twelve years of grappling with the specification of that general idea have considerably changed my perspective on the issue but left the overall idea intact. This book is the attempt to present it in a coherent form.

The book would not have been possible without the insightful comments, questions and corrections by many friends and colleagues over the years. A very important role in the genesis of the book was played by the Institute Vienna Circle Colloquium (past and present members), where I presented and discussed many of its stages. I want to thank in particular Christian Damböck, Christian Fermüller, Johannes Hafner, Manfred Kohlbach, Daniel Kuby, Christoph Limbeck Lilienau, Miles McLeod and Matthias Neuber for many enlightening discussions. My special thanks go to Richard Nickl for his unwavering insistence on the grand scheme. Of particular importance for the clarification of a number of concepts used in Part I of the book was the cooperation with Stephan Hartmann and Jan Sprenger on a Bayesian formalization of the no alternatives argument (see Section 3.5). I am also highly indebted to friends and colleagues who read preliminary versions of the manuscript or parts thereof and whose excellent comments have substantially improved the book. My thanks go to Jeff Barrett, Christian Damböck, Manfred Kohlbach, Christoph Limbeck Lilienau, Keizo Matsubara, Richard Nickl, John Norton, Fritz Stadler, Kyle Stanford and Karim Thebault. I am also grateful for the very helpful comments of the reviewers of Cambridge University Press. Finally, there are all those with whom I discussed topics and texts which eventually entered the book in some form. Without pretensions of completeness, I want to thank Paolo Aschieri,

Michael Baumgartner, Lorenzo Casini, Delphine Chapuis-Schmitz, Isabelle Drouet, Mehmet Elgin, Tobias Fox, Nikita Golovko, Daniel Grumiller, Reiner Hedrich, Herbert Hrachovec, Martin Kusch, Karl Landsteiner, Luca Moretti, Elisabeth Nemeth, Robert Nola, Ed Slovik, Michael Stöltzner, Derek Turner and Jim Woodward.

I would also like to thank Nick Gibbons and Lindsay Barnes, my editors at Cambridge University Press, for their encouraging and constructive support throughout the editing process. I am grateful as well to the Austrian Science Fund (FWF) (P22811-G17) whose funding provided the framework for writing the book.

The book contains modified versions of material that has been published in a number of scientific articles. I am grateful to the respective journals for permitting the publication of parts of those works. Specifically, material from the following scientific articles has been used in the following chapters: "Underdetermination and theory succession from the perspective of string theory," *Philosophy of Science* **73**(3), 298–322, in Chapters 1 and 3; "Scientific realism in the age of string theory," *Physics and Philosophy* **11**, 1–32, in Chapters 1, 2 and 3; "On the conflicting assessments of the current status of string theory," *Philosophy of Science* **76**(5), 984–996, in Chapter 1; "High energy physics and the marginalization of the phenomena," *Manuscrito* **33**(1), special issue *Issues in the Philosophy of Physics*, 165–206, in Chapters 4 and 5; and "String theory and theory assessment," *Foundations of Physics*, **43**(1), 81–100, in Chapter 6.

Last but not least, I want to thank my wife Lisi for her always charming support. Though we first met quite some time ago, she knows me only in the state of planning or writing this book. It seems time to open a new page.

Vienna,
July 30, 2012

Introduction

Fundamental physics at the beginning of the twenty-first century is faced with a confusing situation. On the one hand, physicists sense that for the first time in the long and successful history of their discipline they may have a realistic chance of fulfilling its most ambitious and longstanding goals. The two large complexes of physical reasoning, micro-physics and cosmology, finally seem to offer hope for a genuine unification within one overall physical conception. String theory, in conjunction with new theoretical concepts in cosmology, has revealed the contours of a truly universal theory of all fundamental aspects of the physical world. On the other hand, many physicists believe that fundamental physics is facing a serious crisis. This crisis seems to be generated by a conjunction of two problems. First, physics is losing contact with empirical testing. While microphysics was driven by a continuous stream of empirical data throughout most of the twentieth century, recent decades have witnessed the increasing importance of theories whose characteristic predictions lie far beyond the reach of contemporary experimental testing. The control of the theoretical evolution by empirical data, a crucial element of the natural sciences, thus appears in danger. Moreover, the attempts to come up with a consistent fully universal theory have led to a staggering increase of mathematical complexity, resulting in a wide spectrum of speculative and empirically unconfirmed, in some cases maybe empirically unconfirmable, new hypotheses. Examples range from the spatial extra-dimensions and higher-dimensional objects of string theory to the prediction of many unobservable universes in cosmic inflation. To make things worse, the analysis of the mathematical structures implied by those hypotheses seems to transcend the capabilities of present-day physicists and mathematicians to an extent that

1

renders a completion of the corresponding theories unlikely for the foresee-able future.[1]

While both described aspects of contemporary fundamental physics, the positive as well as the negative one, are in principle acknowledged by all physicists and knowledgeable observers, strong disagreements have arisen with regard to their significance. Many string physicists and modern cosmologists emphasize an optimistic perspective based on the promises provided by the theories they are working on. The problematic aspects of the present situation to them appear as complications of the kind that have always inhibited scientific progress, just to be overcome at the end by improved methods and more ingenious thinking. Quite to the contrary, many physicists working in other research fields feel that the decoupling from experimental testing virtually destroys the relevance of optimistic claims based on string physics and related theories. These claims in their eyes shrink to delusory hopes fed by a dangerous surrender of scientific restraint.

The present book aims at offering a philosophical interpretation of the ambivalent situation characterized above. It will be argued that the criteria of theory assessment have been significantly transformed in fundamental physics in recent decades. Conceptual characteristics of an individual theory as well as characteristics of the research context within which the theory evolves play an increasingly powerful role in assessing the theory's status and viability. While this does not affect the role of empirical data as the ultimate judge of a theory's viability, it substantially raises the status a theory can acquire in the absence of empirical confirmation. It would be too simple to discredit this development as a deviation from solid scientific reasoning and reject it on those grounds. Such a move would mean to petrify the scientific method at one point in time while ignoring that it has always been volatile and in the end must be seen as the product of the successes and failures of scientific reasoning throughout its history. A serious analysis of the indicated changes therefore must aim at identifying the conceptual basis for the new aspects of scientific reasoning and search for the reasons why this altered conceptual basis has emerged. In the end, the specification of what is acceptable as valid reasoning in a scientific field must be up to the involved scientists. A philosophical analysis like the one to be carried out in this book can have the role of providing a wider conceptual framework within which some of the arguments deployed by scientists may appear in a clearer light.

---

[1] Behind the described problem, and independent from it, still looms a second defining problem of fundamental physics, the old but unresolved foundational problem of quantum mechanics. The present book will not address the latter, however.

Of crucial importance for our analysis is the concept of the underdetermination of scientific theory building by the available empirical data. This "scientific underdetermination," as I shall call it, will be argued to constitute an implicit but crucial object of investigation in contemporary fundamental physics. The most important reasons for the trust scientists have in some theories which lack empirical confirmation can be understood in terms of arguments that scientific underdetermination in the given context is severely constrained. The increasing reliance on those argumentative strategies leads to a substantial empowerment of non-empirical theory assessment. Part I of the book aims at identifying the structure of arguments which infer limitations to scientific underdetermination in the context of string theory. While these arguments can be called non-empirical in a certain sense, they are nevertheless rooted in observation and may be understood in terms of an extension of the conventional horizon of observational input.

The important role played by non-empirical theory assessment in the context of string theory, inflationary cosmology and other parts of contemporary fundamental physics seems disturbing to many observers because that kind of reasoning has no place in the canonical understanding of scientific progress that has emerged over the last two centuries. Part II of the book demonstrates that the actual significance of non-empirical theory assessment in physics, and specifically the significance of assessments of limitations to scientific underdetermination, has always been much higher than conceded by the standard philosophical reconstructions of the scientific process. The empirical confirmation of microphysical objects implicitly rests on argumentative strategies of the very same kind as those deployed in non-empirical theory assessment. From the early empirical confirmation of the atom to the current measurements of the Higgs particle, elements of non-empirical theory assessment are necessary for taking claims of the empirical discovery of microphysical objects seriously. The very strong position of non-empirical theory assessment in theories like string physics in this light appears as the strengthening of an already well-established element of scientific reasoning.

Once one accepts the scientific significance of assessments of limitations to underdetermination, a second striking feature of string theory becomes more easily comprehensible. By many of its exponents, string theory is understood to be a candidate for a final theory, a theory that can, at a fundamental level, account for all physical phenomena observable in our world. It is a genuinely philosophical question whether a final theory claim can make epistemological sense at all. Many scientific observers deny this and take the final theory claims put forward in the context of string theory to indicate the over-optimistic mind-set prevalent among string physicists. It is argued in Part III that a universal rejection of final theory claims as epistemologically unjustified is based pre-cisely on the canonical perspective on theory assessment that fails to account for the scientific role of assessments of underdetermination. Once taken into account, assessments of underdetermination can provide a foundation for mean-ingful final theory claims. Assessments of scientific underdetermination thus have the potential to alter our view of scientific theories in two ways. They can push our understanding of a theory's viability beyond the limits of what is currently empirically testable and can provide a basis for understanding the absolute position of current scientific theories within the framework of possible scientific theory building. The latter point has a significant impact on the scientific realism debate, which is discussed in the final chapter of this book.

The present book does not constitute an attempt to demonstrate at a philo-sophical level the viability of string theory or the related final theory claims. Both questions are of a genuinely scientific nature and must be evaluated by applying the apparatus of scientific reasoning to the emerging empirical and non-empirical evidence. The book's core message may be formulated in the following way. The novelty of current theories in fundamental physics is not confined to the conceptual level of those theories themselves. Rather, it extends to the meta-level of theory assessment where a shift of the balance between empirical and theoretical elements can be observed. That shift influences how contemporary physicists in high energy physics and cosmology see their theo-ries. Beyond that, it may eventually alter the philosophical understanding of the relation between a physical theory and the world.

PART I

Delimiting the unconceived

An observer looking at elementary particle physics at the beginning of the 1970s would have found a fairly optimistic scientific community that felt at the brink of a substantially improved understanding of microphysics. Previous decades had produced a steady influx of novel and often confusing empirical data that seemed to require new ideas beyond the well-established techniques of quantum field theory and quantum electrodynamics. In answer to that data, a series of new theoretical conceptions had been developed and had sharpened the understanding of the consistency problems which stood between the status quo and a satisfactory theory of all nuclear interactions. The emerging new theory, later to be called the standard model of particle physics, seemed to be the first convincing candidate for a coherent description of all nuclear interactions and was quickly establishing a new framework of thinking about microphysics. It made a wide range of empirical predictions which awaited empirical testing. Physicists could reasonably expect the next decade to decide the theory's fate.

Forty years later, we find fundamental physics in a more ambivalent mood. Recent decades have impressively fulfilled the expectations physicists were having in the 1970s. The standard model has indeed been vindicated experimentally and has led to a remarkable sequence of consistent predictive success. The last standard model prediction found empirical confirmation in the summer of 2012 at the Large Hadron Collider (LHC) experiments at CERN and thereby concluded an important phase in the evolution of fundamental physics. Moreover, theoretical progress has continued far beyond the standard model and led to a number of far-reaching new theories. Grand unified theories (GUTs) conjecture a more unified structure of nuclear interactions. Supersymmetry (SUSY) posits a more extensive symmetry structure that connects particles of different spin. Supergravity extends that concept towards a theory of gravity. String theory has been developed as the first powerful and promising candidate for a unified theory of all interactions.

Finally, inflationary cosmology substantially altered our perspective on the early universe and brought a rapprochement of cosmological model building and high energy physics. All mentioned theories which reach out beyond the standard model share one common problem, however. Though each of them was first formulated several decades ago, none of them has found empirical confirmation up to now. The canonical experimental strategy of testing ever higher energies by building ever larger particle colliders is becoming increasingly difficult to sustain due to the enormous efforts required for raising the energy levels of those machines. The huge LHC experiment at CERN might be the last experiment of its kind. Supersymmetry is the only main theory that reaches out beyond the standard model and may have good chances of getting empirical confirmation during the LHC experiments. All other theories have characteristic energy levels which must be expected to lie far beyond the range of feasible collider experiments. Cosmic inflation shows some promising consistency with cosmological data, but its basic tenets look difficult to confirm conclusively. The other mentioned theories have little hope of getting significant empirical confirmation in the foreseeable future at all. The increasing detachment of theory building from empirical confirmation may be taken – and indeed is taken by many – as the dawn of a serious crisis of fundamental physics. The shining example of microphysical progress from early atomic physics to the confirmation of the standard model demonstrates how strongly scientific success depends on the close interaction between theory building and empirical confirmation. Once that connection gets too loose, scientific progress may be suspected to slow down significantly or even come to a halt.

Nevertheless, a number of characteristics of contemporary high energy physics and cosmology suggest a more optimistic picture. First and foremost, the theoretical development after the advent of the standard model has provided concrete perspectives for a genuine unification of all known interactions. Thus, the evolution of physics arguably has made significant progress towards a goal that constituted a distant focal point of physical research ever since the times of Newton: the construction of a truly universal physical theory. Moreover, a look at the dynamics of theory building over the last 40 years shows a situation that differs in several respects from what one might expect in a scientific field that has spent those 40 years without empirical guidance. A lack of empirical data might be expected to lead to a proliferation of fundamentally different scientific approaches whose merits cannot be conclusively tested as long as the empirical dearth continues. The scientific community might split up into small groups adhering to those various approaches. The actual situation in particle physics shows a very different picture, however. One observes a high degree of directedness and uniqueness of theory-building at the most fundamental conceptual level. In a number of contexts, one preeminent theory dominates research and is being developed consistently by a large community of physicists over several decades.[1] Directly related to this focus and directedness of theory building, scientists often have a high degree of trust in their theories despite the lack of empirical confirmation. String theory and cosmic inflation are the prime examples in this respect.

The dearth of empirical data, though clearly constituting an obstacle to scientific progress, thus seems to get alleviated to some extent by characteristics of the research process. One may try to explain this situation by purely sociological means, e.g. by assuming that the involved physicists share some ingrained tendency to stick to the ideas of their group and believe its theories. The alternative explanation, which shall be investigated in the following, is based on the understanding that the conceptual environment in which the theories are developed in some way favors the observed directedness of theory building. If that were the case, the same mechanisms which are responsible for the directedness of theory building could also indicate the viability of the corresponding theories and thus constitute the basis for the trust physicists have in contemporary theories despite the absence of empirical confirmation. The following analysis looks for mechanisms of the described kind. It will be argued that such mechanisms indeed exist and incur a substantial shift of the strategies of theory assessment in contemporary fundamental physics.

One particular physical theory may be called the prime example of the situation characterized in the previous paragraphs. String theory aims at giving a unified description of all physical interactions and thereby provides the most far-reaching perspective on a universal understanding of fundamental physics. It may be called the conceptual center of contemporary fundamental physics that connects many other important theories. It is arguably more detached from empirical testing than any other current theory in fundamental physics. At the same time, trust in string physics among its exponents is particularly high. The theory thus provides the most adequate basis for the ensuing analysis of the structure of non-empirical theory assessment.

---

[1] At the level of specific model building, to the contrary, one often finds the very kind of proliferation of theoretical models one would expect in the absence of empirical data. The difference between the vast spectrum of models, that is of possible specific realizations of the physical principles which define a fundamental theory, and the conspicuous lack of alternatives at the most fundamental level of theory building is a striking feature of the present situation in high energy physics.

1
String theory

1.1 A brief introduction to string theory

The evolution of fundamental physics can be construed as a series of unifications. Its beginnings can be traced back to Newton's introduction of a universal gravitational force that provided a unified explanation of celestial phenomena and gravitational phenomena on earth. About two centuries later, Maxwell developed a unified description of light, electric and magnetic phenomena. In 1905, Einstein's special relativity provided a coherent framework for classical mechanics and electrodynamics. A decade later, general relativity expanded this new perspective, making it compatible with the phenomenon of gravity. After quantum mechanics had opened a new world of microphysics ruled by the principles of Heisenberg's uncertainty and quantum statistics in the 1920s (which itself may count as the exception to our rule since it was motivated by accounting for new phenomenology rather than by a quest for unification), quantum physics was soon made compatible with special relativity by the introduction of quantum field theory. In the 1960s and early 1970s, the standard model of particle physics made another step towards unification: a specific form of internal symmetries, gauge symmetry, provided a basis for a coherent description of all three nuclear forces that had been discovered in nuclear and particle physics.

In the 1970s, there remained one fundamental obstacle to an overall description of all known fundamental physical phenomena: the theories of nuclear interactions, which were based on the principles of quantum physics, stubbornly resisted all attempts to be reconciled with general relativity. It became increasingly clear that the standard framework of quantum fields did not allow for any satisfying solution of this problem. Something completely new was needed. The idea that stepped up to play this role was string theory.[1]

---

[1] The topical standard work on string theory is Polchinski (1998). The classic book on the foundations of string theory is Green, Schwarz and Witten (1987). A more easily accessible

String theory was first proposed as a universal theory of microphysics in 1974 (Scherk and Schwarz, 1974).[2] The approach had to struggle with big conceptual difficulties in the beginning. For a long time, it was not clear whether string theory met the most basic requirements for providing a theory of matter. In 1984, Green and Schwarz (1984) finally succeeded in writing down a coherent Lagrangian of a quantized string that included matter fields (the so-called superstring). From that time onwards, string theory has constituted the most prominent and influential attempt to formulate a universal theory of all known interactions. String theory builds on the conceptual foundations that have been established in elementary particle physics in the 1970s. It is a quantum theory that aims at reproducing the interaction and symmetry structure of a gauge field theory. Within this framework, the basic idea of string theory is a fairly simple one: the point-like elementary particles of traditional particle theories are replaced by one-dimensional strings.

In order to understand why this looks like a promising step towards providing a basis for the unification of quantum physics and gravitation, we have to say a few words about the core obstacle to an integration of gravity in the context of quantum field theory: the non-renormalizability of quantum gravity.[3] The calculation of a scattering process in quantum field theory is based on a perturbation expansion that sums up all possible patterns of particles being emitted and absorbed in the process. These possible patterns are represented by the so-called Feynman diagrams which can be calculated. In the calculation of Feynman diagrams one encounters infinite terms which, roughly speaking, arise due to the possibility of point particles coming arbitrarily close to each other. Once we have infinite terms in our calculation, however, we risk losing the capacity of making meaningful quantitative predictions. In gauge field theories, this problem is solved based on the technique of renormalization: the infinities can be ejected from all phenomenologically relevant quantitative results by introducing a finite number of counter-terms to the infinite terms that arise in the calculation. In other words, all ratios between observable quantities have well-defined finite values because the infinities which arise in the calculations cancel

---

textbook is Zwiebach (2004). More recent books are Becker, Becker and Schwarz (2006) and Ibanez and Uranga (2012). A popular presentation for the non-physicist that gives an instructive picture is Greene (1999). The early history of string theory is told by its main exponents in Capelli, Castellani, Colomo and Di Veccia (2012). Early philosophical texts on string theory are Weingard (1989), Butterfield and Isham (2001) and Hedrich (2007a, 2007b).

[2] The history of the concept of strings even goes back to the late 1960s, when it was discussed as a candidate for a description of strong interactions (Veneziano, 1968). Only after it had turned out to fail in that context did it find its new purpose as a universal theory.

[3] A good survey of topical approaches to quantum gravitation can be found in Murugan, Weltman and Ellis (2012). A collection of philosophical papers on the topic can be found in Callender and Huggett (2001).

each other in a controlled way. If one includes gravity in a gauge field theoretical description, however, the renormalization program fails. This failure is related to the fact that the gravitational force grows linearly with increasing energies of the interaction process. As a consequence, the infinities which arise in a perturbative expansion of the gravitational interaction process are particularly "dangerous" and it is no longer possible to cancel all of them by a finite number of counter-terms.

Calculations can still be carried out by introducing an energy cut-off (i.e. integrating energies only up to a certain energy scale). As the result then depends on the cut-off value, however, this approach must be based on the assumption that a new, more fundamental theory that is renormalizable or finite can explain the choice of the cut-off scale. A gauge field theoretical approach to quantum gravity thus can work as an effective theory but cannot serve as a fundamental theory. String theory is understood to offer a solution to the problem of intractable infinities in quantum gravity. The extendedness of the strings "smears out" the contact point between any two objects and thus provides a decisively improved framework that seems to allow finite calculations.[4]

The introduction of extended elementary objects that leads up to string theory thus is chosen for entirely theoretical reasons, in order to provide a coherent unification of the particle physics research program with gravity. So far, no immediate empirical signatures of the extendedness of elementary objects have been observed. We therefore know that, if string theory is a viable theory at all, the string length must be too small to be measurable by current experiments. Since the string length is quite directly related to the gravitational constant, the most natural expectation would be that the string scale lies quite close to the Planck scale, the scale where the gravitational coupling constant (which grows linearly with the interaction energy) becomes of order one. If that was the case, canonical scenarios would imply that the string scale lies about 13 to 14 orders of magnitude beyond the so-called electroweak scale,[5] that is the energy scale testable by the current LHC experiment at CERN. Under some specific circumstances, however, which will be discussed a little later, the string length might lie much closer to empirical observability.

String theory relies on the core principles of a perturbative expansion of relativistic quantum mechanical interaction processes and offers a conceptual modification that seems capable of solving the coherence problems of that

---

[4] Though the finiteness of string theory is not proven conclusively, it is supported by fairly strong evidence.
[5] The electroweak scale corresponds to the masses of the heaviest particles of standard model physics. Technically, it is the scale where the electroweak symmetry is spontaneously broken (see Section 4.1).

approach in the context of gravitational interaction. In this sense, string theory represents a natural continuation of the high energy physics research program. It turns out, however, that the seemingly innocent step from point-like objects to strings has a wide range of complex structural consequences which lead far beyond the conceptual framework of point-like high energy physics.

A first important implication can be derived directly from the quantization of the string. It turns out that a coherent (i.e. anomaly free) quantization of the string is possible only if the string is moving in a spacetime with a specific number of dimensions. In particular, a string theory that can describe matter particles (and not just photons and other particles of integer spin) can only be consistently formulated in ten spacetime dimensions. This prediction marks the first time in the history of physics that the number of spatial dimensions can be derived in a physical theory. The obvious fact that only four spacetime dimensions are macroscopically visible can be taken into account by the assumption that six dimensions are "compactified": they have the topological shape of a cylinder surface – or, if more than one dimension is compactified, of a higher-dimensional torus – where, after some translation in a "compactified" direction, one ends up again at the point of departure.[6] Just like the string length, the compactification radius must be assumed to be so small that the additional dimensions are invisible to the current experiments in high energy physics and gravitational physics. The canonical picture would put the compactification radii more or less at the string length and close to the Planck length. There do exist theoretical scenarios of large or "warped" extra dimensions, however, where Planck scale and string scale merely give the misleading impression of lying many orders of magnitude beyond the electroweak scale as long as one does not account for the propagation of gravity through the large or warped extra dimensions.[7] In such scenarios, the string scale could be low enough even for becoming observable at the LHC.

---

[6] Technically, the space of compact dimensions is described by a Calabi–Yau manifold.

[7] Those scenarios are based on models where some of the extra dimensions are transgressed only by gravitation while nuclear interactions are bound to the other spatial dimensions. Large extra dimensions which are only transgressed by gravity (Antoniadis, Arkani-Hamed, Dimopoulos and Dvali, 1998) can be close to the micrometer range without contradicting present experiments because precision measurements of gravity have not yet been carried out below that scale. If gravity radiated off into large extra dimensions, it would get strongly diluted. The effective four-dimensional gravitational constant we observe would therefore be very small even if the higher-dimensional gravitational coupling were rather strong. From our four-dimensional perspective this would create the impression of a large scale difference between the electroweak scale and the Planck scale. Warped dimensions (Randall and Sundrum, 1999) are characterized by a peculiar geometry that 'thins out' propagation through those dimensions. Models of warped extra dimensions explain the weakness of the gravitational coupling by utilizing this "thinning out" effect. A warped extra dimension can have the effect that gravitons are very unlikely to be found close to the lower dimensional subspace to which nuclear interactions are confined. This amounts to a suppression of the effective gravitational constant on that subspace.

In conventional quantum physics, elementary particles carry quantum numbers which determine their behavior. A particle's characteristics like spin or charge, which are expressed by quantum numbers, constitute intrinsic and irreducible properties. Strings, to the contrary, do not have quantum numbers. They can differ from each other only by their topological shape and their dynamics. Strings can be open, meaning that they have two endpoints, or closed like a rubber band. If they are closed, they can be wrapped around the various compactified dimensions in different ways. Both open and closed strings can assume different oscillation modes. These characteristics determine the macroscopic appearance of the string. To the observer lacking experimental tools of sufficient resolution for perceiving the stringy structure, a string in a specific oscillation mode and topological position looks like a point-like particle with certain quantum numbers. A change of the oscillation mode of the string would be perceived as a transmutation into a different particle. Strings at a fundamental level do not have coupling constants either. The strength of their interaction with each other again can be reduced to some aspect of their dynamics. (The ground state of a certain mode of the string expansion, the dilaton, gives the string coupling constant.) All characteristic numbers of a quantum field theory are thus dissolved into geometry and dynamics of an oscillating string. It turns out that the string necessarily contains an oscillation mode that corresponds to the graviton (a massless spin 2 particle). Therefore, string theory automatically includes gravity. String theories which are able to describe matter fields are more difficult to formulate. They have to be supersymmetric, i.e. they must be invariant under specific transformations between particles of different spin. String theoretical models which have this property are called "superstring" models.

A fundamental problem faced by string physicists may be characterized the following way. String theory has started from a perturbative point of view. As noted above, a perturbative approach to a relativistic quantum theory describes interaction processes by expanding them as a series of Feynman diagrams. These Feynman diagrams can in principle be arbitrarily complex and involve an arbitrarily high number of particle exchanges. In a case like electroweak interaction, we know from experiment that the coupling constant is small so that Feynman diagrams with high numbers of particle exchanges (which are suppressed by a high factor of the coupling constants) can be neglected in approximate calculations.[8] Perturbation theory has thus proved to be a highly powerful technique in the context of weakly coupled quantum field theory.

---

[8] In fact, no rigorous mathematical proof of the convergence of perturbation theory in quantum field theory, and therefore no proof of the viability of perturbation theory to all orders, has been found so far. We just know from comparisons of perturbative calculations with experimental data that an expansion up to some rather low order in the expansion parameter provides a good approximation to the overall theory.

In string physics, the situation is more difficult than in quantum field theoretical descriptions of nuclear interactions. Neither the overall theory behind the perturbative expansion nor the size of the expansion parameter are known. In fact, as we have mentioned above, the string coupling, which constitutes the expansion parameter of a perturbative expansion of string theory, is no fundamental free parameter but itself emerges from the dynamics of the fundamental theory. String theorists thus cannot trust perturbative calculations. They have to look for non-perturbative information about the theory in order to acquire an understanding of the theory's general characteristics. Some limited progress has been made in this direction.

An important feature of string physics which has shed some light on the nonperturbative structure of string theory is the occurrence of string dualities. The string world shows a remarkable tendency to link seemingly different string scenarios by so-called duality relations. Two dual theories or models are exactly equivalent concerning their observational signatures, though they are constructed quite differently and may involve different types of elementary objects and different topological scenarios. An example of a duality relation that conveys the basic idea of duality in a nice way is T-duality. String theory, as has been mentioned above, suggests the existence of compactified dimensions. Closed strings can be wrapped around compactified dimensions like a closed rubber band around a cylinder and they can move along compactified dimensions. Due to the basic principles of quantum mechanics, momenta along closed dimensions can only assume certain discrete quantized eigenvalues. Thus, two basic discrete numbers exist which characterize the state of a closed string in a compactified dimension: the number of times the string is wrapped around this dimension, and the eigenvalue of its momentum state in that very same dimension.[9] Now, T-duality asserts that a model where a string with characteristic length[10] l is wrapped n times around a dimension with radius R and has momentum eigenvalue m is dual to a model where a string is wrapped m times around a dimension with radius l^2/R and has momentum eigenvalue n. The two descriptions give identical physics.

T-duality is not the only duality relation encountered in string theory. The existence of dualities turns out to be one of string theory's most characteristic features. Duality relations are conjectured to connect all different types of superstring theories. Before 1995, physicists knew five different possible types of superstring theory which differed by their symmetry structure and therefore seemed physically different. Then it turned out that these five string

---

[9] The two numbers are called "winding number" and "Kaluza–Klein level," respectively.
[10] The characteristic string length denotes its length when no energy is being invested to stretch it.

theories[11] and a sixth by then unknown theory named "M-theory" were interconnected by duality relations (Witten, 1995; Horava and Witten, 1996). Two kinds of duality are involved in this context. Some string theories can be transformed into each other through the inversion of a compactification radius, which is the phenomenon just discussed under the name of T-duality. Others can be transformed into each other by inversion of the string coupling constant. This duality is called S-duality. Then there is M-theory, which can be reached from specific types of string theory[12] by transforming the coupling constant into an additional 11th dimension whose size is proportional to the coupling strength of the corresponding string theory. M-theory contains two-dimensional membranes rather than one-dimensional strings.[13] Despite their different appearances, duality implies that the five types of superstring theories and M-theory only represent different formulations of one single actual theory. This statement constitutes the basis for string theory's uniqueness claims and shows the pivotal role played by the duality principle.

Another important duality relation is the AdS/CFT correspondence that also goes under the name holographic principle. String theory (that is, a theory that contains gravity) in a space with a specific geometry (so-called anti-de Sitter space, in short AdS) is conjectured to be empirically equivalent to a pure supersymmetric gauge theory (which is a theory that contains no gravity) on the boundary of that space (that is, on a space whose dimension is reduced by one compared to the AdS space). In other words, a duality relation is conjectured to hold between a theory with gravity and another one without gravity. Each gravitational process that takes place in AdS space can be translated into a corresponding non-gravitational process on the boundary space.

The AdS/CFT correspondence has recently led to highly fruitful applications which lie far beyond the limits of string theory proper and do not address the question how to unify all physical interactions (Policastro, Son and Starinets, 2001; Kovtun, Son and Starinets, 2004). It has turned out that the AdS/CFT correspondence can be used in contexts of complex QCD-calculations (calculations of scattering processes involving strong interactions) where all conventional methods had failed. In that case, the physical situation can be described by a gauge theory and

---

[11] Generally, the term string theory denotes the overall theory that describes the five types of string theory and their relations to each other. However, string physicists often address the individual types of string theory in short as individual string theories as well. Though this is slightly misleading – in particular in the light of their connectedness by dualities – we will stick to that manner of speaking. It should become clear from the context when the term string theory is used in the general sense and when in its more limited sense.

[12] Type I and type IIA string theory.

[13] M-theory is insufficiently understood at this point. It may have the potential to lead towards a more fundamental understanding of string theory.

AdS/CFT correspondence leads from there to a gravitational theory that can be calculated more easily. Though strictly speaking the investigated systems cannot be described by a gauge theory that has a gravitational dual (they are not super-symmetric and involve massive fermions), calculations in the dual picture in many cases turn out to provide significantly better results than more conventional methods. The successes of these calculations have recently led to the spread of string-based methods far beyond the borders of the theory itself. The present book will not be concerned with those investigations any further. They must be mentioned, however, as a striking example of the wide range of string theoretical reasoning in present-day physics.

Coming back to genuine string theory, the analysis of non-perturbative aspects of string physics based on dualities has led to important new insights. It was understood that the spectrum of physical objects in string theory was far wider than initially expected. Beyond the initially posited one-dimensional elementary objects, consistency required the additional introduction of a spectrum of various higher-dimensional objects. These objects are called D-branes, where an added number can denote the number of spatial dimensions. (A D5-brane, to give an example, is an object with five spatial dimensions.)

In recent years it has been better understood how to construct string theory ground states with stable compact dimensions based on combined systems of D-branes and fluxes.[14] It turns out that the freedom for constructing such states is huge. Estimates regarding the number of possible ground states go up to 10^{500} or even higher. Working on the question of how to deal with the variety of ground states – which is often called the string landscape (Susskind, 2003) – is one important task for string theory today. One intensely discussed suggestion is to turn the large number of ground states from a problem into a blessing and use it for explaining the fine-tuning of the cosmological constant.[15]

String theory lies at the core of many important developments in high energy physics and cosmology today. Supersymmetry, a highly influential theory that is currently being searched for at the LHC experiments at CERN, is predicted by string theory (though not necessarily at energy scales accessible to the LHC) and constitutes one of the theory's core characteristics. Supergravity, which for some time was seen as a promising candidate theory for a unification of nuclear interactions and gravity, today is mostly deployed as an effective theory of string theory. Discussions of grand unified theories, another highly influential theory in high energy physics, today are often led within a string theoretical

---

[14] Fluxes are oscillation modes of the string which do not correspond to particles known from point-like particle physics.

[15] See Section 7.5.

framework. Eternal inflation, the leading theory on the early evolution of the universe, is closely linked to string theory with regard to the theory's basic layout as well as with regard to more far-reaching considerations based on the anthropic principle.[16] One can speak of an interrelated web of theories in contemporary high energy physics and cosmology that is held together by the conception of string theory.

The current theoretical state of development of string theory may be characterized the following way. Four decades of intense work on the theory carried out by large numbers of theoretical physicists and mathematicians have not resulted in the construction of a complete theory. String theory today constitutes a complex web of reasoning consisting of elements of rigorous mathematical analysis, of general conjectures which are based on reasoning in certain limiting cases, of modeling that is done within specified frameworks and of some approximate quantitative assessments. The resulting understanding provides a vast body of structural information and theoretical interconnections between various parts of the theory but leaves unanswered many crucial questions. No real breakthrough has been achieved that would allow specific quantitative calculations of observables from the fundamental principles of string theory. The tediously slow progress witnessed by string physicists over the last few decades does not raise expectations of finding a completion of string theory in the foreseeable future. It seems more realistic to expect a continuation of the kind of development that has characterized the theory's evolution so far: a sequence of small steps of theoretical progress, interspersed with some significant conceptual breakthroughs and periods of slowed down theory dynamics.

Empirically, the situation is similarly complicated. String theory has not found empirical confirmation up to now. In order to understand the chances for future empirical tests, it is important to distinguish between the theory's core characteristics and its implications for physics at lower energy scales. The core property of strings, their extendedness, is expected to show only close to the Planck length. In classical scenarios, that means that the extendedness of the string becomes empirically testable only about 13 orders of magnitude[17] beyond the energy scales that can be reached by the most powerful high energy experiment today, the LHC experiments at CERN. There is no hope to reach those scales within the framework of collider physics. We have mentioned above, however, that some scenarios imply a far lower Planck and string scale and thus move both of them towards regions that might be testable by collider physics. A second core prediction of string physics is the existence of extra dimensions. An empirical

---

[16] A little more on the theories mentioned in this paragraph will be said in Section 4.2.
[17] Which corresponds to a factor of 10 trillion.

discovery of extra dimensions could not be taken as full confirmation of string theory since extra dimensions might also occur in non-stringy scenarios. However, they are not implied by any contemporary theory other than string theory. If a prediction as "esoteric" as extra dimensions turned out to be vindicated by experiment, this discovery would clearly be considered strong corroborative evidence for string theory. The empirical perspectives for the discovery of extra dimensions are similar to those for the discovery of extended elementary objects. The classical scenarios put the size of extra dimensions about 13 orders of magnitude beyond the reach of the LHC. Some scenarios of large or warped extra dimensions would imply, however, that large extra dimensions could be testable by precision measurements of gravity at short distances or by the LHC. To conclude, there is no clear indication that core predictions of string theory can be tested by experiments conceivable today. However, under certain specific conditions, such empirical confirmation might be possible.

String theory may also be supported based on predictions regarding physics at the electroweak scale. If empirically measured parameter values of standard model physics which are not implied by the standard model itself or its GUT or SUSY extensions turned out to be predicted by string theory, this would be taken to constitute strong corroboration of string theory. Depending on the range of such successful predictions, they could assume the status of outright empirical confirmation of string theory even in the absence of direct evidence for extended strings. To consider the most extreme example, let us assume that all standard model parameter values turned out to be precisely predicted by string theory. In that case, it would seem very plausible to believe in string theory's validity even without having ever observed an extended string. At this point, the chances that any specific predictions of low energy parameter values could be derived from string physics are difficult to assess. While the fundamental structure of string theory is understood to be determined uniquely based on very general pre-assumptions, the current understanding of string physics suggests that a vast number of string theory ground states can be constructed from string physics. The selection of the ground state "we are living in" determines the values of all those parameters which define our observable world. Since this selection constitutes the outcome of a quantum process, its prediction must remain of a statistical nature, just like the outcome of some individual microphysical process. This implies a considerable reduction of string theory's predictive power. The present incomplete understanding of string theory does not allow a clear assessment as to what extent string theory retains predictive power under the stated conditions. Though speculations that string physics might end up predictively empty seem hardly tenable based on the current physical understanding, the unclear situation naturally adds to the impression that string physics is unsatisfactorily detached from empirical confirmation.

This "statistical" problem is superimposed by the problem of the insufficient understanding of string theory dynamics. As of today, it is not possible to derive any quantitative predictions from basic principles of string physics. Therefore, even to the extent that string theory may predict low energy parameter values, string physicists today are not able to specify and calculate those predictions. The most promising strategy for making some contact with observation may consist in trying to analyze to what extent some general properties of physics at the electroweak scale are implied by or seem likely in the context of string physics. If found, coherence of that kind between string theory and the observed world could provide some degree of corroboration for string physics.

Given its unsatisfactory theoretical state and the lack of empirical confirmation, string theory clearly must be called an unconfirmed speculative hypothesis according to the canonical paradigm of theory assessment. This does not square well, however, with the theory's actual status in the field of high energy physics. String theory has attained a pivotal role in fundamental physics and has been treated as a well-established and authoritative theory for quite some time by the community of string theorists and by physicists in related fields. As we have described above, large parts of fundamental physics are influenced by string theoretical analysis. The string community is one of the largest communities in all of theoretical physics and for many years has produced the majority of the field's top-cited papers. Moreover, many string theorists express a remarkably strong trust in their theory's viability. Though they certainly acknowledge their theory's theoretical incompleteness and the lack of empirical evidence for it as deplorable obstacles, most of them believe that the theoretical quality of string theory in itself justifies the claim that the theory constitutes an important step towards a deeper understanding of nature. The serious mismatch between the status one would have to attribute to string theory based on the canonical paradigm of theory assessment and the status the theory actually enjoys is being reflected by the intense dispute that has arisen about the status of string physics in recent years. The next section will have a closer look at that discussion.

1.2 The conflicting assessments of the current status of string theory

Before entering into the details of the dispute about the status of string theory, it is important to clarify the role that dispute is going to play in the context of this ook. Primarily, the book aims to explain the mechanisms of theory assessment in string theory and related theories. The dispute among physicists that will be discussed in this section is not essential to that analysis. If the dispute had not arisen, the philosophical motivation for developing the arguments presented in this book would have remained unaffected and the core of those arguments would have remained unchanged. However, the dispute may be understood as an additional indicator that something philosophically interesting is happening in physics today that is capable of creating serious divides within the physics community at a deep conceptual level. In other words, the dispute can serve as a marker that theoretical physicists currently face a situation where philosophical considerations on the conceptual foundations of their ways of reasoning can be of interest to them. On that basis, the ensuing analysis of the dispute can serve as a test case for the philosophical perspective suggested in this book. If, as I hope to be able to convey, the suggested perspective is capable of providing a convincing explanation of the dispute, it may be taken to be supported by this explanatory success.

So let us take a closer look at the conflict. Soon after the theoretical break-through of string theory in 1984, some degree of skepticism developed among physicists in other fields with respect to the string theorists' high level of trust in their theory. This skepticism grew with time, as string theory remained empiri-cally unconfirmed and theoretically incomplete whereas its exponents showed no sign of abandoning their strong self confidence. In recent years, criticism of string physics has been presented in an increasing number of articles and books, which made the conflict about the status of string physics clearly visible to a wider public. Let me illustrate the irreconcilably different points of view by citing four remarkably different statements on string physics.

> During the last 30 years of his life, Albert Einstein sought relentlessly for a so-called unified field theory – a theory capable of describing nature's forces within a single, all-encompassing, coherent framework. [...] Einstein never realized his dream [...]. But during the past half-century, physicists of each new generation – through fits and starts, and diversions down blind alleys – have been building steadily on the discoveries of their predecessors to piece together an ever fuller understanding of how the universe works. And now long after Einstein articulated his quest for a unified theory but came up empty-handed, physicists believe they have finally found a framework for stitching these insights together into a seamless whole – a single theory that, in principle, is capable of describing all phenomena. The theory [is] string theory.
> *Brian Greene,* The Elegant Universe *(1999, p. IX)*

> The moment you encounter string theory and realize that almost all of the major developments in physics over the last hundred years emerge – and emerge with such elegance – from such a simple starting point, you realize that this incredibly compelling theory is in a class of its own.
> *Michael Green (1997), quoted in Greene,* The Elegant Universe *(1999, p. 139)*

Imagine a tourist trying to locate a specific building in a vast and completely unfamiliar city. There are no street names (or at least none that make any sense to the tourist), no maps and no indication from the totally overcast sky as to which directions are north, south or whatever. Every so often there is a fork in the road. Should the tourist turn right or left, or perhaps try that attractive little passageway hidden over to one side? The turns are frequently not right angles, and the roads are hardly ever straight. Occasionally, the road is a dead end street, so steps must be retraced and another turning made. Sometimes a route might then be spotted that had not been noticed before. There is no-one around to ask the way; in any case, the local language is an unfamiliar one. At least the tourist knows that the building that is sought has a particular sublime elegance, with a supremely beautiful garden. That, after all, is one of the main reasons for looking for it. And some of the streets that the tourist chooses have a more obvious aesthetic appeal than the others. [. . .] Each successive choice of turn is a gamble, and on frequent occasions you may perhaps feel that a different one held more promise than the one [. . .] actually chosen. [. . .]

If there are too many of these ["choices"], the chance of guessing right each time may become exceedingly small.
> *Roger Penrose on string physics' chances of success in* The Road to Reality
> *(2005, p. 888f)*

Despite a number of tantalizing conjectures, there is no evidence that string theory can solve several of the big problems in theoretical physics. Those who believe the conjectures find themselves in a very different intellectual universe from those who insist on believing only what the actual evidence supports. The very fact that such a vast difference of views persists in a legitimate field of science is in itself an indication that something is badly amiss.
> *Lee Smolin,* The Trouble with Physics *(2006, p. 198)*

While the four citations may be particularly outspoken, they do represent the two main positions which pertain among physicists with regard to the current status of string physics. On one side of the divide stand most of those physicists who work on string physics and in fields like inflationary cosmology[18] or high energy particle physics model building, which are strongly influenced by string physics. That group represents a slight majority of physicists in theoretical high energy physics today. Based on an internal assessment of string theory and the history of its development, they are convinced that string theory constitutes a crucial step towards a better and more genuine understanding of the world we observe. On the other side stand many theoretical physicists of other fields, most experimental physicists and most philosophers of physics. They consider string theory a vastly overrated speculation.

We witness a confrontation of two sharply diverging positions without much appreciation of the respective opponent's arguments. It is no big exaggeration to

---

[18] A little more about inflationary cosmology will be said in Section 4.2.

say with Smolin that the string theorists and the critics cited above, though they are all theoretical physicists, live in different worlds. Remarkably, they do so despite the fact that they largely agree on the problems string theory faces. All problems discussed in the previous section with regard to the chances of string theory making contact with empirical data are acknowledged fully by string physicists as well as critics of string theory. The differences between the two sides lie in the conclusions drawn.[19]

In the dispute described, the string critics play the role of defenders of the classical empirical paradigm of theory assessment. Therefore, it makes sense to start the characterization of the two diverging positions by looking at their perspective. The string critics' case shall be presented largely based on Lee Smolin's book (Smolin, 2006) and Roger Penrose's remarks on the topic in Penrose (2005). Similar arguments can be found e.g. in Woit (2006) and Hedrich (2007a, 2007b). By and large, the sketched arguments are representational of the considerations which have led many physicists who do not work in string physics or particle physics model building towards adopting a skeptical assessment of the current status of string physics.

Penrose and Smolin base their assessments on their canonical understanding of the scientific process: scientific theories must face continuous empirical testing in order to avoid going astray. As formulated suggestively in Penrose's citation above, the steady sequence of theoretical alternatives which open up in the course of the evolution of a research program makes it seem highly implausible that the scientific community could consistently make the right theoretical choices in the absence of empirical guidance. If empirical testing remains absent for a long period of time, the chances seem high that scientists will find themselves – to use Penrose's picture – lost in a wrong part of town. Therefore, in order to be conducive to scientific progress, scientific theories are expected to fulfil a certain pattern of evolution. A theory is expected to reach a largely complete theoretical state within a reasonable period of time. Only after having reached a fairly complete state does a theory allow for a full assessment of its internal consistency and can it provide quantitative predictions

---

[19] It should be mentioned at this point that Lee Smolin, one of the few exponents of the string critical camp with working experience in string physics, does differ substantially from other string physicists in his scientific assessment of string theory's structure as well. In particular, he doubts the viability of most of those mathematical conjectures which constitute the backbone of string physics. Regarding these arguments, the dispute can be understood as a conventional example of the occurrence of different opinions within a scientific field. In this narrower context, however, the divergent position of one individual scientist would be of limited interest to philosophy of science and would not suffice to motivate the fundamental debate that arose in recent years. Therefore, the present discussion will leave aside Smolin's internal string theoretical assessments and stick to those arguments which drive the debate at a more general level.

of empirical data. The theory then can be expected to undergo empirical testing within a limited time frame in order to decide whether further work along the lines suggested by the theory makes sense or, in the case the theory was empirically false, would be a waste of time.

Looking at the present condition of string theory about four decades after it was first proposed as a fundamental theory of all interactions, we can see that it has achieved neither of the goals described in the last paragraph. Even in the eyes of most of its critics, this does not imply that the theory should be fully abandoned. It has happened before that theories have taken too long to reach maturity and empirical testability and have been shelved until, after some period of general experimental or theoretical progress, they turned out to have the capacity of contributing to scientific progress after all. The critics' point is, however, that one cannot know whether or not such late success will occur before a theory has found empirical confirmation. A theory that has not reached theoretical completion and empirical confirmation therefore cannot be called successful by classical scientific standards. Given the large number of physicists who have worked on string theory with high intensity over the last 30 years, the theory may actually be called remarkably unsuccessful by those standards. Thus, its critics take string theory as a scientific speculation that may deserve a certain degree of attention due to its interesting theoretical properties but is unfit to play the role of a pivotal, let alone dominating, conceptual focal point of an entire scientific discipline. Still, this is exactly what string theory has been doing for more than a quarter of a century now.

Smolin and Penrose criticize string theorists for ignoring the canonical rationale for theory assessment that was presented above and for developing an unwarranted degree of trust in their theory's validity. According to Smolin, string theorists systematically overestimate their theory's "performance" by creating their own criteria of success which are tailored to be met by string physics. Examples of this strategy would be the straightforward interpretation of mathematical progress as physical progress without empirical backing or the string theorists' frequent allusions to structural beauty (see e.g. Michael Green's citation in this section). Smolin argues that such "soft" criteria create arbitrary mirages of genuine scientific success. Their application in his eyes impedes the field's ability to carry out an objective assessment of its progress and moves the field away from legitimate scientific reasoning with respect to theory appraisal. The resulting overestimation of the theory's status, according to Smolin, disturbs a healthy scientific process since it binds to string physics too many resources whose allocation to other parts of physics could produce more significant results. It is important to emphasize that the thrust of the string-critical arguments questioning the scientific viability of string physics focuses on the strategies of theory evaluation deployed by string physicists. It does not target the methods applied in the construction of string theory, the scientific quality of which remains largely uncontested.

The question as to how a considerable share of the most eminent physicists can jointly commit the described serious methodological blunder is answered by Smolin at a sociological level by deploying the concept of groupthink. The latter phenomenon allegedly tends to occur in professional groups with high status, strong internal competition and intense internal interaction. Under such circumstances, the members of a group may be forced into the unreflected adoption of the group's standard positions by a mix of intellectual group pressure, admiration for the group's leading figures and the understanding that fundamental dissent would harm career perspectives. An all too positive and uncritical self-assessment of the group is the natural consequence.

Let us now turn to the string theorists' perspective. No string theorist would deny the problems the theory faces. But most of them believe there to be strong theoretical reasons for placing trust in the theory's viability despite these problems. String theorists adopt an altered understanding of the balance between empirical and theoretical methods of theory appraisal that amounts to a massive strengthening of the status of non-empirical theory assessment in the absence of empirical confirmation.

Before entering the analysis of the conceptual reasons for the described shift, I want to address an interesting general aspect of the view string theorists have of their critics. As described above, critics of string theory take string theorists to violate principles of canonical scientific behavior by having undue trust in an empirically unconfirmed theory. String theorists have a similar kind of complaint about many critics of string theory. They tend to consider the disregard shown by many non-string physicists for their theory-based reasons to believe in string theory a blatant violation of the scientific expert principle. The latter principle establishes an informal code of mutual respect for scientific specialization among scientists in different fields, which is taken to be conducive to an optimized appreciation of the overall body of scientific knowledge by scientists and external observers. According to this principle, non-experts are expected to base their opinion about the content and status of theories in a well-established scientific field largely on the assessments by those who have established themselves as experts in the field. In the case of string physics, this principle often seems to be discarded. Most critics of string theory from other fields base their criticism largely on their own general scientific intuition, bolstered by the statements of a couple of prominent opponents of string theory who do have expertise in the field but are not amongst the field's foremost figures. Exponents of string theory tend to locate the reasons for this unusual phenomenon in a lack of openness and flexibility with respect to the acceptance of new and unusual physical ideas on the side of the string critics.[20]

The debate between string physicists and critics of string theory on the legitimate assessment of the theory's success and viability thus is characterized to a significant extent by allusions to psychological or sociological aspects which are alleged to impede a sober scientific assessment on the side of the respective opponent. While the present book does not aim at offering a psychological or sociological study of modern fundamental physics, it seems appropriate to make a few comments on the plausibility of the involved arguments at that level.

It is probably fair to say that both sides do have a point. The community of string physicists indeed may be characterized by high status, close interaction among its members, and a research dynamics that is largely guided by a few key figures (this may actually be less true in 2012 than in the 1990s). Presumably, few in the community would deny that aspects of groupthink can be identified in the community's patterns of interaction and behavior. Equally, it is difficult to deny the point stressed by many string theorists that the antagonism between string theorists and physicists who are critical of string physics shows the characteristics of a dispute between the exponents of a new way of thinking and more conservative scientists who are less open to fundamentally new developments and prefer staying closer to known territory.

Still, it appears questionable whether psychological and sociological arguments of the described kind can on their own offer a satisfactory explanation of the dispute about the status of string theory. Regarding the groupthink argument, it may be pointed out that one could very well detect elements of groupthink at several stages of the evolution of theoretical physics throughout the twentieth century. The founding period of quantum physics or the emergence of gauge field theory, to give just two examples, arguably show similar sociological constellations with tightly knit high status groups of physicists who believed in work on a revolutionary new theory, strong leading figures who largely determined the course of events and many followers who just wanted to be part of the game. There are other examples of theories which were supported

---

[20] It should be emphasized that physicists on both sides of the divide are aware of the slightly precarious character of the "non-physical" arguments deployed in the debate. Lee Smolin has applied the concept of groupthink to the community of string physicists (which, incidentally, seems a quite accurate representation of what many critics of string physics do think about string physicists) but is careful not to present it as a core argument. String theorists, when entering a discussion with their critics (see e.g. Polchinski in his reasoning against Smolin (Polchinski, 2007)), try to keep the debate at an entirely physical level. Still, it seems important to mention the sentiment behind the dispute that is not so much expressed in written form as it is in private communication.

by leading figures of physics but still were considered mere speculations due to the lack of empirical confirmation. Throughout twentieth century physics, the scientific method has proved strong enough to keep scientific control mechanisms intact despite the presence of elements of groupthink. A convincing groupthink argument thus would have to explain why it is just now that groupthink has succeeded in sustaining an irrationally positive assessment of a theory for more than a quarter of a century.

Moreover, describing the string community solely in terms of groupthink would be grossly one-sided and inadequate. Many of today's most creative and innovative physicists work in the field. The string community is arguably characterized by a particularly high degree of openness for new ideas and a marked tendency to question old ways of reasoning. This is reflected by the large number of new ideas and ideas coming from other fields of physical research which have been adopted in string physics. The picture of a sheepish group following the directives of a few prophets would be an obvious misrepresentation of the actual situation in the field.

On the other hand, the string theorists' allusions to their opponents' backwardness does not provide a fully convincing explanation of the emergence of the string-critical point of view either. A number of strong critics of string physics have contributed important and innovative physical concepts themselves (Roger Penrose and Lee Smolin both are examples). It seems implausible to relate their string criticism to a generally conservative scientific attitude. As an overall phenomenon, the degree to which physicists working in other fields refuse to adopt the string theorists' assessment of their theory and thereby implicitly distance themselves from the expert principle is as striking as it is atypical in twentieth century physics.[21] It seems to require an explanation that goes beyond a simple attestation of scientific conservativism.

Irrespective of the actual explanatory power of the involved scientists' allusions to psychological and sociological arguments, it is significant, however, that such arguments are addressed by physicists at all. Differences of opinion in physics usually tend to be discussed by evaluating the argumentative content of the opposing position. A retreat to criticism at a psychological level may be taken to suggest that the opponents are no longer able to reconstruct the opposing position in a rational way. Such a situation can arise when the two sides enter the discussion with incompatible predispositions which are themselves not sufficiently addressed in the dispute. In the following, it shall be argued that this is exactly what is happening in the dispute on the status of string theory.

---

[21] Comparable developments did occur during the initial stages of relativity and quantum physics but soon degenerated into pseudo-scientific fringe phenomena.

Let us reconstruct the argument between string physicists and string critics in a little more detail. String theorists have built up considerable trust in their theory based on the theory's internal characteristics and the history of its evolution. Critics of string theory protest that the string theorists' trust in their theory is not tenable on the basis of generally acknowledged scientific criteria of theory assessment. String theorists retort that the convincing quality of string theory, being based on the theory's specific structural characteristics, only reveals itself to the string theory expert, which implies that most of the critics are just not competent to evaluate the situation. The critics, in turn, do not feel impressed by this argument, since, according to their own understanding, they make a general point about the character of the scientific process, which should remain unaffected by any specific analysis of the theory's technical details.

The dispute can be construed as a discussion that fails to be productive due to a paradigmatic rift between the two disputants: each side bases its arguments on a different set of fundamental preconceptions. This paradigmatic rift, however, differs from the classical Kuhnian case in two respects.

First, it is placed at a different conceptual level. In Kuhn's picture, paradigmatic differences can be identified by looking directly at the involved scientific theories. To give an example, Newtonian physics represents the paradigm of deterministic causation while quantum mechanics introduces a new paradigm that allows for irreducible stochastic elements in the dynamics of physical objects. Paradigmatic shifts of this kind may have far-reaching implications at all levels of the understanding of the scientific process. Still, they are rooted in and implied by conceptual differences between the corresponding theories themselves. The paradigmatic rift between proponents and critics of string theory is of a different kind. It cannot be extracted directly from conceptual differences between specific scientific theories but only arises at the meta-level of defining the notion of viable scientific argumentation. In the dispute, the critics of string theory mostly do not contradict claims of string theory itself but question the strategies of theory assessment applied in the context of string physics. One could thus call the rift between string theorists and their critics "meta-paradigmatic" in the sense that it cannot be discussed without focussing on the meta-level question of the choice of viable criteria of scientific theory assessment.

Second, the development of the new paradigm did not happen in a revolutionary step. Rather, the understanding of theory assessment evolved gradually based on the scientific experiences of scientists in the field. Only once that gradual process had lasted for some time and scientists then looked outside the limits of their field, did they discover the paradigmatic rift that had opened up between their understanding and the canonical paradigm of theory assessment prevalent in other parts of physics.

A mutual understanding between string theorists and their critics is prevented by the fact that the meta-paradigmatic character of the rift between the two sides is not addressed as such in the discussion. While string theorists proudly point out the conceptual novelties of their theory, they tend not to emphasize the meta-paradigmatic shift at the level of theory assessment that was induced by the evolution of string theory. A number of reasons may be responsible for this fact. First, being physicists rather than philosophers, string theorists naturally focus on their theory's direct physical import and consider the functionality of the scientific process a pre-condition that is more or less taken for granted. Second, since the meta-paradigmatic changes evolved gradually for the scientists in the field, they look less dramatic to them than to outside observers. And third, conceding a deviation from canonical scientific praxis would invite a level of criticism that string physicists have no interest to incur.

The critics of string theory, on the other hand, develop their arguments without acknowledging the possibility that a shift of the scientific paradigm might constitute a scientifically legitimate development under some circumstances. Therefore, they discuss string physics strictly based on the canonical scientific paradigm and interpret each mismatch between the two straightforwardly in terms of string theory's failure to meet scientific standards.

Both sides thus agree in disregarding the meta-paradigmatic aspect of their discussion. In doing so, however, they actually lead the discussion based on incompatible sets of hidden preconceptions and therefore must miss each other's point. Seen from either perspective, the respective opponent's position does not have legitimacy based on the preconceptions taken to provide the valid framework for the entire debate. The recourse to mutual imputations of personal insufficiencies follows as a natural consequence. Claims of scientific hubris thus are summoned against claims of insufficient intellectual acuteness.[22]

---

[22] A recent paper by Johannsen and Matsubara (2011) attempts an assessment of the current status of string theory while explicitly denying the "meta-paradigmatic" character of the changes associated with the evolution of string theory. Thereby, Johannsen and Matsubara run into a problem that is closely related to the one described above. They discuss the status of string theory within the framework of the Lakatossian concept of research programmes (Lakatos, 1970) and try to determine whether string theory has constituted a progressive or a degenerative research program in recent decades. Applying the canonical criteria of theory assessment, they come to the conclusion that string theory constitutes a degenerative research program: for three decades it has been developed without generating new and specific empirical predictions. As they do acknowledge, however, this flies in the face of the string theorists' own assessment of their research program. Most string physicists would consider their research program progressive due to the new theoretical insights they have won on its basis. Seen from the perspective presented in this book, there is a clear reason for the difficulties faced by Johannsen and Matsubara: by denying the meta-paradigmatic character of the shifts related to string theory, they forsake the freedom of modifying the criteria for the progressive character of the theory. Thus they end up with an assessment that cannot do justice to the arguments presented by the scientists in the field.

Framing the controversy in terms of a shift of the scientific paradigm allows acknowledgement of the reasonability of both positions on the basis of their respective preconceptions. Largely unintentionally, string physicists have been led towards a novel conception of scientific theory appraisal by their scientific research, which they had carried out in accordance with all standards of scientific reasoning. The scientific process itself thus has led beyond the canonical limits of scientific reasoning. The rise of the new understanding of theory evaluation was clearly accelerated by the fact that the stronger role of purely theoretical argumentation it suggested came in handy in view of the lack of available empirical support for string theory. The lack of empirical support was not the primary source of the former development, though. As will be shown in the next sections, the theory itself and the circumstances of its evolution provide substantial reasons for the altered perspective. Those reasons, however, remain invisible to physicists in other fields who have not experienced the scientific dynamics that instigated the described shift of the scientific paradigm. Therefore, they must understand the described shift as a purely defensive ad hoc measure instigated by the long empirical drought and see no sound scientific reason to follow it. Within the framework of their traditional and well-tested scientific paradigm, they find ample justification for repudiating the string theorists' assessments of string theory's status.

Is the string theorists' move a legitimate one? Can it be legitimate if good arguments support it within the string theorists' own framework of reasoning? Clearly, a shift of the scientific paradigm that is induced by the dynamical evolution of a research field must be considered a legitimate option in principle. No paradigm of scientific reasoning has been installed as a god-given law before the commencement of scientific research. Rather, such paradigms have emerged based on the successes and the failures of scientific reasoning witnessed by scientists in the past. Novel scientific input thus must be expected to alter the scientific paradigm in the future. The question whether an emerging shift of the scientific paradigm actually constitutes an improvement over the prior situation can be very difficult to answer, however.

In one respect, characterizing the dispute between string physicists and their critics in terms of a paradigmatic rift seems particularly apt. To a greater extent than in most cases of scientific theory change, the debate on string physics is affected by fundamental difficulties to decide between the two positions on an "objective" basis, i.e. without anticipating the outcome by employing the preconceptions related to one or the other position. Since the difference in the understanding of what counts as scientific success is crucial to the difference between the two paradigms, it is obviously impossible to decide straightforwardly between the two paradigms by assessing scientific success. Both paradigms allow for plausible criticism of the respective opponent on their own grounds. Seen from the critic's perspective, it is quite plausible to argue that a modification of the scientific paradigm in times of crisis is counter-productive, as it carries the risk of overlooking a solution that would satisfy the criteria set up by the old paradigm. Seen from the perspective of the new paradigm, sticking to the old criteria too long inhibits scientific progress by sticking to a misguided chimera of the static nature of scientific principles. All one can do in this case is assess the internal coherence and attractiveness of the old and new positions on their own terms and compare the two internal assessments on a more general – and therefore necessarily more vague – argumentative basis.

I argued in Section 1.1 that the traditional paradigm of theory assessment has run into a substantial crisis in the context of present-day fundamental physics. The next chapters will have a closer look at the question whether and to what extent the newly emerging paradigm of theory assessment in string physics and some other fields can provide a viable basis for overcoming that crisis. Let us begin by looking at the conceptual reasons string theorists rely on for believing in their theory. Later, an attempt will be made to put those reasons into a philosophical perspective.

1.3 Three contextual arguments for the viability of string theory

Why do string physicists invest trust in the viability of their theory? The main problem in this respect is addressed clearly in Penrose's cited text. Penrose raises one fundamental worry with regard to overly confident assessments of theories that have not been empirically confirmed: those assessments are always threatened by the possibility that other scientific explanations of the available data have been overlooked so far. Kyle Stanford has called this general problem of scientific reasoning the problem of unconceived alternatives (Stanford, 2001, 2006). Any reliable assessment of a theory's status on theoretical grounds must answer Penrose's worry by addressing the question of unconceived alternatives in some way. It shall be argued in the following that the current assessments of the status of string physics rely on a number of arguments which indeed amount to addressing that question.

Roughly, string theorists rely on two basic kinds of arguments when developing trust in their theory. Arguments which address structural characteristics of string theory itself shall be discussed in Part III of this book. Part I will focus on the other group of arguments, the contextual arguments which are based on general characteristics of the research process that leads towards string theory.

These arguments do not rely on any specific properties of the theory itself and are therefore of wider relevance. Part II will be devoted to demonstrating that they constitute an important part of theory assessment in fundamental physics in general.

Three main contextual reasons for the trust string theorists have in their theory may be distinguished. While all three arguments are "common lore" among string physicists, it is difficult to pinpoint a "locus classicus" for each of them. One can find a combination of all arguments in Chapter 1 of Polchinski (1998) and in Polchinski (1999). The first and third argument appear in Greene (1999, Chapter 1).

The plain no alternatives argument (NAA): string theorists tend to believe that their theory is the only viable option for constructing a unified theory of elementary particle interactions and gravity. It is important to understand the scope and the limits of that claim. String theory is not the only theory dealing with questions of quantum gravity. Various forms of canonical quantum gravity try to reconcile gravity with the elementary principles of quantum mechanics. They discuss the question of unification at an entirely different level than string theory, however. The latter stands in the tradition of the standard model of particle physics and is based on pivotal concepts such as non-abelian gauge theory, spontaneous symmetry breaking and renormalizability.[23] The goal of string theory is to reconcile gravity with these advanced and successful concepts of contemporary particle physics and therefore to provide a truly unified description of all natural forces. In this endeavor, the traditional investigations of canonical quantum gravity do not constitute alternatives, which leaves string theory as the only available way to go.[24] That is not to deny the relevance of the investigations of canonical quantum gravity. String theorists would just argue that, once the viable results of canonical quantum gravity are put into the context of contemporary particle physics, they will blend into the string theory research program.

Why is it so difficult to find a unified description of gravity and nuclear interactions? We have encountered the crucial problem for a unification of point-particle physics and gravity already in Section 1.1: quantum gravity is non-renormalizable within the traditional field-theoretical framework.

---

[23] A little more about those concepts will be said in Part II.

[24] There also exists a tradition of thought that questions the necessity of quantizing gravitation in a theory that gives a coherent description of quantum physics and gravitation. (Recent works are Wüthrich (2004), and Mattingly (2005).) Though some ideas concerning quantum theories of gravity without quantized gravity have been put forward, as yet none of them has been formulated in any detail, however. Like canonical quantum gravitation, those considerations at the present point address the reconciliation of gravity with basic quantum physics but do not offer concepts for a coherent integration of gravity and advanced particle physics.

Non-renormalizable quantum gravity cannot be considered viable at the Planck scale, however, the scale where the gravitational coupling becomes strong. Early attempts to solve this problem applied the traditional methods of gauge field theory and tried to deploy symmetries to cancel the infinities which arise in loop calculations and therefore make the theory finite. For some time, the concept of supergravity, which utilizes supersymmetry, looked like a promising candidate for carrying out this task, but eventually the appeal to symmetry principles was judged insufficient.[25] As it turns out, the remaining theoretical options are quite limited. One might venture into giving up some of the most fundamental pillars of present-day physics like causality or unitarity. Ideas in these directions have been considered, but did not lead to any convincing theoretical schemes. If one wants to retain these most fundamental principles, then, according to a wide consensus, there remains only one way to go: drop the idea of point particles, which univocally leads to string theory (see e.g. Polchinski, 1998, 1999).

A body of explicit analysis supports the notion that there may be no alternatives to string theory. Those considerations can be exemplified by an argument in Polchinski (1999). Polchinski starts with an innocent looking posit of a position–position uncertainty relation instead of the posit of extended elementary objects. He shows that the efforts to make that idea work eventually imply the very string theory he had set out to circumvent. Based on a number of arguments of a similar kind in conjunction with the fact that no alternatives have come up despite intense search, it may be suggested that string theory is the only option for finding a unification of all interactions within the framework of the long standing fundamental principles of physics.[26]

Naturally, the claim of alternatives does not remain uncontested. On what basis can we be confident that the scientists' lack of alternative ideas is not just a consequence of their limited creativity? Who can rule out that one of the most

---

[25] The case has not been closed entirely until this day. Recent analysis has suggested that there might be a finite ($N = 8$) supergravity after all (Bern, Dixon and Roiban, 2006). It still seems doubtful for a number of reasons, however, whether supergravity can be a fully consistent theory on its own.

[26] The history of science contains earlier claims of univocal inference from observation to the theoretical scheme. Newton's "deduction from the phenomena" has been taken up in Norton (1993, 1994), while Worrall (2000) has emphasized that deduction's dependence on prior assumptions. Newton's claim is based on the assertion of an immediate and intuitively comprehensible connection between observation and theoretical explanation. Compared to the case of Newton, the situation in string theory has decidedly shifted towards the assertion of a radical limitedness of options for mathematically consistent theory construction while the intuitive aspect of Newton's argument has been dropped. Whether this shift, in itself, enhances the authority of the string theoretical claim of no choice may be a matter of dispute. Part III of this book will demonstrate, however, that the string theoretical claim can be embedded in an entirely new and more powerful argumentative framework.

fundamental principles of physics indeed has to be jettisoned at this stage to describe nature correctly and that string theory is nothing more than a delusive "easy" way out that just does not accord with nature? Indeed, candidates for alternatives do sometimes appear where no alternatives had been in sight before and thus pose a constant threat to simple arguments of no alternatives.[27] It seems necessary to look for additional arguments for the viability of string theory which, rather than focussing on the scientists' difficulties to think of alternatives, are related to qualities of the theory itself.

Probably the most important argument of that kind is **the argument of unexpected explanatory coherence (UEA)**. It is widely held that a truly convincing confirmation of a scientific theory must be based on those of the theory's achievements which had not been foreseen at the time of its construction. Normally, this refers to empirical predictions which are later confirmed by experiment. However, there is an alternative. Sometimes, the introduction of a new theoretical principle surprisingly provides a more coherent theoretical picture after the principle's theoretical implications have been more fully understood. This kind of theoretical corroboration plays an important role in the case of string theory. Once the basic postulate of string physics has been stated, one observes a long sequence of unexpected deeper explanations of seemingly unconnected facts or theoretical concepts. Let us have a brief look at the most important examples.

String theory posits nothing more than the extendedness of elementary particles. The initial motivation for suggesting it as a fundamental theory of all interactions was to cure the renormalizability problems of quantum field theories that include gravity. Remarkably, string theory does not just provide a promising framework for quantum gravity but actually implies the existence of gravity. The gravitational field necessarily emerges as an oscillation mode of the string. String theory also implies that its low energy effective theory must be a Yang–Mills gauge theory, and it provides the basis for possible explanations of the unification of gauge couplings at the GUT-scale. The posit that was introduced as a means of joining two distinct and fairly complex theories, which had themselves been introduced due to specific empirical evidence, thus turns out not just to join them but to imply them.

---

[27] The case of new arguments for the finiteness of supergravity was mentioned in footnote 25. Another recent example for an interesting new perspective is Horava (2009), which proposes a scenario that solves the renormalizability problem of quantum gravity in a different way. It presents a theory that is non-relativistic at very short distances but approximates relativistic physics at longer distances. If consistent, such a scenario would constitute an alternative to string theory as a possible solution to the renormalizability problem of quantum gravity. The debate on the scenario's consistency, its promises and its limitations is ongoing at this point.

String theory also puts into a coherent perspective the concept of super-symmetry. Initially, interest in this concept was motivated primarily by the abstract mathematical question whether any generalization of the classical continuous symmetry groups was possible. As it turns out, supersymmetry is the maximal consistent solution in this respect. Soon after the construction of the first supersymmetric toy-model, it became clear that the implementation of supersymmetry as a local gauge symmetry (i.e. supergravity) had the potential to provide a coherent quantum field theoretical perspective on gravity and its interaction particle, the graviton. In the context of string theory, on the other hand, it had been realized early on that a string theory that involves fermions must necessarily be locally supersymmetric.[28] The question of the maximal continuous symmetry group, the quest to integrate the graviton naturally into the field theoretical particle structure, and the attempts to formulate a consistent theory of extended elementary objects thus conspicuously blend into one coherent whole.

A problem that arises when general relativity goes quantum is black hole entropy. The necessity to attribute an entropy proportional to the area of its event horizon to the black hole in order to preserve the global viability of the laws of thermodynamics was already understood in the 1970s (Bekenstein, 1973). The area law of black hole entropy was merely an ad hoc posit, however, lacking any deeper structural understanding. In the 1990s it turned out that some special cases of supersymmetric black holes allow for a string theoretical description where the black hole entropy can be understood (producing the right numerical factors) in terms of the number of degrees of freedom of the string theoretical system (Strominger and Vafa, 1996). Thus, string physics provides a structural understanding of black hole entropy.

All of these explanations represent the extendedness of particles as a feature that seems intricately linked with the phenomenon of gravity and much more adequate than the idea of point particles for a coherent overall understanding of the interface between gravity and microscopic interactions. The subtle coher-ence of the implications of the extendedness of elementary objects could not have been foreseen at the time when the principle was first suggested. It would look like a miracle if all these instances of delicate coherence arose in the context of a principle that was entirely misguided.

To what extent is it justified to have serious confidence in the viability of string theory's phenomenological predictions on the basis of the presented

---

[28] World sheet supersymmetry of a string that includes fermions was discovered by Gervais and Sakita (1971). A string theory that shows local target space supersymmetry was finally formulated by Green and Schwarz (1984).

"non-empirical no miracles argument"? There do exist cases in the history of science where the inference from a concept's success to its viability was invalidated by the fact that just one aspect of the concept was responsible for the success while important parts of the concept were misguided. A prominent example would be the ether theories whose success was based on the viability of the wave equation but whose core concept, the ether, had to be dropped eventually. In the case of string theory it is difficult to imagine anything of that kind, since the concept is based on one simple and entirely structural posit which would seem impossible to reduce without taking it back altogether. It must be considered a genuine possibility, however, that a deeper, more fundamental principle than string theory itself could be responsible for unexpected explanatory interconnections without implying string theory itself. Still, as broad a spectrum of unexpected explanatory interconnections as is encountered in the case of string theory, in conjunction with the apparent difficulty to come up with consistent alternatives to the theory, may seem difficult to reconcile with the idea that those explanatory successes do not hinge on the theory itself. One thus may have the impression that the argument of unexpected explanatory interconnections has some strength but may not feel confident enough to rely on it without further analysis. Given that there is no option at this point for carrying out empirical tests of string theory itself, it would thus seem important at least to find a possibility of checking the general validity of the non-empirical arguments for string theory's viability on an empirical basis. The third argument is based exactly on this kind of empirical test at a meta-level.

The meta-inductive argument from the success of other theories in the research program (MIA): most string theorists, at any rate those of the first generation, are mainly educated in traditional particle physics. Their scientific perspective is based on the tremendous predictive success of the particle physics standard model. The latter was created for solving technical problems related to the structuring of the available empirical data (in particular, the problem of making nuclear interactions renormalizable) and it predicted a whole new world of new particle phenomena without initially having direct empirical confirmation. Just like in the case of string theory, it turned out that none of the alternatives to the standard model that physicists could think of was satisfactory at a theoretical level. In addition, surprising explanatory interconnections emerged. (For example, the distinction between a confining interaction like strong interaction and the non-confining electromagnetic interaction could be explained as a natural consequence of the difference between a non-abelian and an abelian interaction structure.) Given the entirely theoretical motives for its creation, the lack of satisfactory alternatives and the emergence of unexpected explanatory interconnections, the standard model can be called a direct precursor of string theory.

Indeed, string theorists view their own endeavor as a natural continuation of the successful particle physics research program. The fact that the standard model theory was at the end impressively confirmed by experiment conveys a specific message to particle physicists: if you knock on all doors you can think of and precisely one of them opens, the chances are good that you are on the right track. Scientists working on unifying gauge field theory and gravity have thought about all currently conceivable options, including those which drop fundamental physical principles. The fact that exactly one approach has gained momentum suggests that the principles of theory selection which have been successfully applied during the development of the standard model are still working.

It is important to emphasize that MIA relies on empirical tests of other theories and thereby in a significant sense resembles the process of theory confirmation by empirical data. The level of reasoning, however, differs from that chosen in the classical case of theory confirmation. In the present context, the empirically testable prediction is placed at the meta-level of the conceptualization of predictive success. We do not test the scientific theory that predicts the collected empirical data but rather a meta-level statement. The following hypothesis is formulated: scientific theories which are developed in the research program of high energy physics in order to solve a substantial conceptual problem, which seem to be without conceptual alternative and which show a significant level of unexpected internal coherence tend to be empirically successful once they can be tested by experiment. This statement is argued for based on past empirical data (in our case, largely data from the standard model of particle physics and from some earlier instances of microphysics) and can be empirically tested by future data whenever any predictions which were extracted from theories in high energy physics along the lines defined above are up to empirical testing.

All data that can be collected within the high energy physics research program thus constitute relevant empirical tests of the viability of MIA and thereby have implications for the status of other theories in the field which are considered likely viable based on that hypothesis. Any experiment that confirms predictions whose viabilities have been considered likely based on purely theoretical reasoning would improve the status of scientific theories of similar status in the research field. On the other hand, the status of non-empirical theory evaluation and therefore also the status of theories believed in on its basis would seriously suffer if predictions that are strongly supported theoretically were empirically refuted. Trust in string physics on that basis is influenced by new empirical data even when that data does not represent a test of string theory itself.

Present-day high energy physics provides two excellent examples for the described mechanism. At the LHC, two theories were and still are tested which, before the start of the experiment, were considered probably viable (to different degrees) based on theoretical reasoning. First, the canonical understanding of high energy physics implied that the LHC was likely to find a Higgs particle.[29] The Higgs mechanism constituted the only known method of producing the observed masses of elementary particles in a gauge theoretical framework. Since gauge field theory had proved highly successful in the standard model, where all empirical predictions other than the Higgs particle had already been empirically confirmed, and since the theoretical context of mass creation by spontaneous symmetry breaking[30] in gauge field theory was well understood, physicists would have been profoundly surprised if the Higgs particle had not been found at the LHC. (A closer analysis of that case will be carried out in Chapter 4.) The actual discovery of the Higgs particle in summer 2012 clearly constitutes an example of an eventual empirical confirmation of a theory that had been conjectured and taken to be probably viable on theoretical grounds. Thereby, it constitutes a confirmation not just of the Higgs theory but also of the meta-level hypothesis that the involved theoretical strategies of theory assessment are viable. In the given case, the general belief in the existence of the Higgs particle was so strong that its discovery did not alter overall perspectives too much. A failure to find the Higgs, however, would have constituted a serious blow not just to the current understanding of the standard model but to the status of non-empirical theory assessment in general. String physicists in that case would have had to answer the question on what basis they could be confident to get the basic idea of string theory right on entirely theoretical grounds if high energy physics could not even correctly predict the Higgs particle. Since the principle of the viability of non-empirical theory assessment can only be of a statistical nature, it could not be refuted by individual counter instances. Trust in string physics would have seriously suffered, however, if a prediction as well trusted as the one regarding the Higgs particle had failed.

The second theory that may get empirically confirmed at the LHC is low energy supersymmetry. The situation there is a little different than in the case of the Higgs particle. As will be discussed in more detail in Section 4.2, some conceptual arguments do hint towards low energy supersymmetry. The cogency of those arguments, however, is less clear than in the Higgs case or in the case of string theory. Many string physicists would argue that there remain more conceptual options for avoiding low energy supersymmetry than for avoiding

---

[29] Either as a fundamental or as a composite particle.
[30] See Section 4.1 for a brief explanation of spontaneous symmetry breaking.

string theory. Thus, if low energy supersymmetry was not found at the LHC that would obviously not help the trust in string theory but the damage would be limited. On the other hand, if both the Higgs particle and low energy supersymmetry were found, that would be taken as significant support for the reliability of theory generation based on conceptual reasoning. Trust in non-empirical theory assessment would get considerably strengthened, which in turn would enhance the trust in string theory.[31]

Having discussed the three most important arguments of non-empirical theory assessment, it is now possible to be more specific about what is meant by "non-empirical" in the given context. The term "non-empirical" clearly does not imply that no observation or no empirical data has entered the argument. As we have seen, "non-empirical" theory assessment does rely on observations about the research process, the performance of scientists looking for alternative theories or the success of theories in the research field. What distinguishes empirical from non-empirical evidence in the given sense is the following. Empirical evidence for a theory consists of data of a kind that can be predicted by the theory assessed on its basis. Non-empirical evidence, to the contrary is evidence of a different kind, which cannot be possibly predicted by the theory in question. No scientific theory can predict that scientists will not find other theories which solve the same scientific problem (the observation that enters NAA). Nor can a scientific theory possibly predict that other theories which are developed within the research field independently from the given theory tend to get empirically confirmed (the observation that enters MIA). Non-empirical evidence for a theory thus is evidence that supports a theory even though the theory does not predict the evidence.

In conjunction, the three presented arguments of non-empirical theory assessment lay the foundations for trust in string theory. All three reasons have precursors in earlier scientific theories but arguably appear in string theory in a particularly strong form. The complete lack of empirical evidence in the given case leads to a situation where non-empirical theory assessment is solely responsible for the status attributed to string theory and therefore is of special importance.

---

[31] The described support for string theory at the meta-level must be distinguished from reasoning at the theoretical "ground level." The discovery of supersymmetry would support string theory at the theoretical ground level as well, since string theory predicts supersymmetry (though it does not imply low energy supersymmetry).

2

The conceptual framework

Section 1.3 presented the reasons for trust in string theory from the perspective of physicists. A number of crucial questions remain open at that level, however. Is there a common basis for the three arguments for having trust in string theory? Can those arguments be called genuine scientific reasoning? Do they amount to theory confirmation? Finally, can we find a philosophical background story behind the described rise of non-empirical theory assessment?

In order to deal with those questions, we have to take a step back and define the philosophical context within which the presented situation evolves. I will begin by offering a more substantial philosophical definition of the canonical paradigm of theory assessment that motivates much of the criticism against string theorists' trust in their theory. Based on that characterization, I will then introduce the core philosophical concept which shall be argued to underlie the ongoing paradigm shift with respect to theory assessment. This concept, which will determine much of the later analysis in this book, goes under the name "assessment of scientific underdetermination."

2.1 The classical paradigm of theory assessment

A minimal and fairly uncontroversial first description of theory assessment in science may be given in the following way. Scientific theories make predictions that can be tested by collecting empirical data. If the collected data turns out to be in agreement with the theory's predictions, this enhances the scientists' trust in the theory's viability (in whatever way we may want to specify that trust in terms of the attribution of truth, empirical adequacy, the reliability of its empirical predictions or other concepts). If the data contradicts the predictions, that lowers trust in the theory correspondingly.

39

Most philosophers of science, in particular those who are guided by the example of physics, take one crucial step beyond the characterization given above. They insist that confirmation by empirical data is the only way a scientific theory can acquire the status of an acknowledged and well-established scientific theory. According to this understanding, scientists are only entitled to believe in their theory if that theory has been empirically confirmed by empirical data. The same basic idea may be expressed in terms of the conception of knowledge: a scientific theory can only constitute knowledge about the world if it has been confirmed by empirical data. Based on the classical definition of knowledge as justified true belief, the statement means that the scientist cannot get justification for her belief in a theory without empirical confirmation. In terms of an alternative definition of knowledge as belief that has been produced by a reliable cognitive process, empirical confirmation constitutes a necessary element of any reliable cognitive process that leads towards establishing a scientific theory.

The exclusive role of empirical testing in scientific theory assessment is evident in the most influential schemas of the scientific process. According to hypothetico-deductivism, scientists produce a new theory based on prior knowledge of empirical data by an act of creative speculation. The resulting theory at first has the status of an untested hypothesis. Empirical predictions are then deduced from that hypothesis. The hypothesis can be tested only by confronting its predictions with actual empirical data. Only if the theory's predictions get repeatedly confirmed empirically, the hypothesis can become a well-established and trusted theory. On the other hand, disagreements between predictions and empirical data weaken and eventually refute the hypothesis. The hypothetico-deductivist thus adamantly holds that confirmation and refutation by empirical data are the only possible ways to determine a theory's viability.

Bayesianism, which currently constitutes the most popular formalization of scientific theory confirmation, conveys a similar albeit not identical message. The (subjective) Bayesian assumes that the scientist attributes a probability of truth to each scientific theory. Bayes' theorem,

$$P(\mathrm{T}|\mathrm{E}) = \frac{P(\mathrm{E}|\mathrm{T})}{P(\mathrm{E})} P(\mathrm{T}),$$

determines how new empirical evidence E alters the probability of the theory's truth. $P(\mathrm{T})$ denotes the prior probability of the truth of hypothesis H before the empirical data E has been considered; $P(\mathrm{E})$ is the probability of the empirical data E disregarding H; $P(\mathrm{H}|\mathrm{E})$ is the probability of H when E has been taken into account; and $P(\mathrm{E}|\mathrm{T})$ is the probability of E given that H is true. Empirical data constitutes confirmation of theory H if

$$P(\text{T}|\text{E}) > P(\text{T}).$$

Probabilities according to this schema can only be assigned based on prior probabilities attributed to theories, which themselves must either be posited or assessed based on another Bayesian procedure. No algorithm exists for determining initial probabilities. They are subjectively chosen by the scientist based on her general assessments, her experience, personal preferences, etc. According to the Bayesian, this irreducible subjective element can be reconciled with the notion of generally agreed upon scientific knowledge because repeated empirical tests tend to create a convergence behavior for probabilities that are derived from a wide range of early prior probabilities. Empirical data that agree with the theory's predictions quickly raise low prior probabilities $P(\text{T})$ to levels close to those reached when starting from rather high $P(\text{T})$s. Empirical data that contradict the theory's predictions, on the other hand, quickly reduce high initial probabilities $P(\text{T})$. Probabilities in mature science are thus taken to be fairly well decoupled from the prior probabilities attributed at early stages of the scientific development.

Unlike hypothetico-deductivism, Bayesianism does not exclude the attribution of very high initial probabilities to scientific theories. Bayesianism is coherent with a situation where the scientist takes a new theory to be very probably true for some reason without having tested it empirically. However, the Bayesian approach strongly emphasizes that the scientific process of theory assessment starts in earnest with the theory's confrontation with empirical data E. Only then do the probabilities that initially have an entirely subjective character, and may be chosen quite differently by individual scientists, converge and thus start constituting reliable scientific judgements. The empirical testing of the theory by new data is what the Bayesian considers the genuinely interesting part of the process of theory assessment.

Bayesianism as well as hypothetico-deductivism implicitly make a second assumption regarding the scientific process. They assume that scientific theories which are to be tested empirically acquire a fairly complete state within a reasonable time span. Only once they have assumed that state is it possible to deduce predictions from the theory along the lines suggested by hypothetico-deductivism; and only sufficiently complete theories can be reasonably evaluated in a Bayesian way. Both conceptions of the scientific process thus distinguish a period of theory construction from a distinct period of empirical testing.

Joining the considerations presented above, we can define the contours of a prevalent overall understanding of the scientific process that is expressed in hypothetico-deductivism as well as in "canonical" Bayesianism. A notion of science along these lines is shared by most philosophers of science today and is taken to be a reliable guide for scientific activity by scientists themselves in modern physics and related fields. The scientific research process is thereby understood in terms of a dichotomy between theoretical conceptualization and empirically based theory assessment. On the one hand, in the realm of theory building scientists create theories in order to structure and describe the available empirical data and, eventually, predict new data and new phenomena; on the other hand, the process of empirical observation and experimentation first inspires theory building and then, based on new data, provides confirmation or refutation of the developed theoretical statements. A theoretical conception has to meet certain structural preconditions in order to count as a candidate for a viable scientific theory: it has to constitute a largely complete and internally coherent theoretical structure; and it has to offer quantitative predictions the empirical side can aim to test. If a conception fulfils those conditions and thus is being awarded a "candidate status" for becoming a scientifically viable theory, empirical testing constitutes the only genuinely scientific method for determining that theory's status. As long as the theory's core predictions have not been empirically confirmed, the theory remains a scientific speculation. Only after empirical confirmation has been forthcoming may the theory be accepted as viable scientific knowledge.

I will call this understanding of the scientific process the classical scientific paradigm. What the classical scientific paradigm unequivocally denies is the possibility that a process of rational analysis on its own can offer an alternative strategy to empirical testing for turning a scientific hypothesis into a well-established and well-trusted theory. Purely theoretical considerations like those pertaining to a theory's simplicity, beauty or apparent cogency may contribute to the scientist's subjective trust in a theory's prospects of being viable. They are acknowledged as being of potential auxiliary value to the researcher who tries to decide which way to go before empirical tests have been carried out. They are not considered an objective and independent factor, however, in determining a given theory's scientific viability.

The assertion of a univocal primacy of empirical data in theory assessment has been questioned and toned down in a number of ways by various philosophers. Two prominent examples may be mentioned.

Thomas S. Kuhn (1962) emphasized that the interpretation of empirical data always happens within some paradigm based on scientific background convictions and thus can never provide an objective account of its own significance. It may happen that aspects of empirical data which are considered important evidence under one paradigm are considered largely irrelevant under another. Kuhn and his followers deny that the confirmation value of empirical data for a theory can be determined on a basis that is independent of the paradigm represented by that theory. This implies that the choice of a scientific paradigm cannot be based on empirical data alone but must be based on a broader form of deliberation whose outcome is not univocally determined by rational analysis and which involves conceptual, theoretical and even social elements.[1]

While Kuhn thus alters the understanding of the role of empirical data in the research process, he does not question that the individual scientist who works within some paradigm (either the one she inherited from her peers or a new one she is testing) looks for confirmation of her theory based on empirical data according to the procedures outlined in previous paragraphs. Though paradigm choice according to Kuhn is not decided based on empirical confirmation or refutation alone, the testing of a theory within its paradigm does proceed based on empirical testing nevertheless. In the eyes of the working scientist, theories thus assume a status of being well established based on empirical testing. In this respect, Kuhn has no quarrel with hypothetico-deductivism.

Larry Laudan's analysis of theory assessment in his book *Progress and its Problems* (Laudan, 1977) moves a little further. Laudan aims at modifying the understanding of theory assessment and theory choice without putting a strong emphasis on the Kuhnian distinction between paradigm change and normal science. Thereby, non-empirical theory assessment turns into a canonical element of theory assessment at all times and is not confined to the revolutionary phases of paradigm change. Laudan argues that the dominating influence of the empiricist paradigm has blinded philosophers of science to the simple fact that scientists judge their theories' status and viability not merely based on empirical merits but also and substantially based on theoretical qualities. What makes this claim significant is Laudan's insistence on the equal status of both strategies of theory assessment. According to Laudan, a careful historical investigation of the process of theory selection in science does not justify asserting a univocal hierarchy between empirical and theoretical strategies of theory assessment. Theoretical strategies do not play a strictly auxiliary role that would restrict their influence to cases when the empirical verdict is not yet clear. Rather, theoretical arguments under certain conditions can be considered more important than empirical ones and may overrule empirically based preferences.

---

[1] Unlike some of his followers, Kuhn himself does not deny that the eventual prevalence of one theory and the demise of its contenders can be rationally explained. According to Kuhn, a period of normal science starts once one paradigm has turned out so successful that its superiority over its known contenders can no longer be rationally denied even from those contenders' perspectives. Only in periods of crisis and paradigm change does the identification of the most successful paradigm actually depend on the paradigm on which the assessment is based (see Hoyningen-Huene, 1993).

The main claim to be pursued in this book shares the basic sentiment behind Laudan's analysis. It shall be argued that the classical paradigm of theory assessment grossly underrates that assessment's theoretical aspect. However, the analysis will choose a significantly different point of departure than Kuhn or Laudan. The conception to be promoted is theory assessment based on assessments of limitations to scientific underdetermination, which does not play a significant role in Kuhn's or Laudan's work. In order to introduce the idea of scientific underdetermination, let us return once more to the foundations of the classical paradigm of theory assessment.

2.2 Scientific underdetermination

Why does it make sense for scientists to believe in the empirical predictions of theories which have already been empirically confirmed and refuse to believe in predictions made by empirically unconfirmed theories? In order to answer that question, we must first introduce the important distinction between prediction based on straightforward induction and prediction of genuinely new phenomena. The first kind of prediction may be exemplified by the expectation that the scattering process between two hard and massive macroscopic objects that has so often been found to proceed in agreement with the Newtonian principles of mechanics will again adhere to those principles the next time we observe it. A classic example of the second kind of prediction would be Poisson's inference from a wave theory of light to a bright spot at the center of a shadow cast on a screen by a round object when hit by light that has passed first through a narrow hole.[2]

Scientists are willing to believe scientific predictions of the first kind because they are willing to rely on the principle of induction. The extension of the applicability of simple enumerative induction to many precisely specified scientific contexts is taken to be one central reason for the success of the scientific principle.

If a scientist constructs a theory, however, that (a) fits the available data and (b) predicts new phenomena which have not yet been observed, her trust in the actual existence of the newly predicted phenomena is restrained by one crucial consideration: other, so far unknown scientific theories may exist which fit the present data equally well but predict different new phenomena. In other words,

---

[2] The borders between the two kinds of predictions cannot be univocally drawn. The two kinds of predictions rather constitute ideal types segmenting a continuous spectrum rather than univocally distinguishable groups. Still, the distinction is helpful for conveying the rationale behind the scientists' dealings with their theories.

Underdetermination by...	(a) all possible evidence:	(b) the available evidence:
(1) Logically:		Hume, problem of induction; Quine-Duhem thesis
(2) Ampliatively:	Quine ('Reasons for Indeterminacy of Translation'); van Fraassen (The Scientific Image')	Sklar, Stanford: transient underdetermination; SCIENTIFIC UNDERDETERMINATION

Figure 2.1 Underdetermination of scientific theory building.

scientific theory building is expected to be significantly underdetermined by the currently available empirical data. I will call this presumption the principle of "scientific underdetermination."

Scientific underdetermination has to be distinguished from two types of underdetermination which figure most prominently in philosophy.[3] In order to distinguish the different kinds of the underdetermination of theory building it may be helpful to draw a systematic picture (see Figure 2.1).

Philosophers of science speak either about (1) underdetermination considering all logical possibilities or about (2) underdetermination under some general assumptions which are taken to be constitutive of all viable scientific research. Such general assumptions, which are called "the ampliative rules of scientific method" in Laudan (1996), may include a commitment to an intuitive form of the principle of induction, some kind of Ockham's razor and the exclusion of non-integrated ad-hoc explanations of individual events. These assumptions may differ from one scientific field to the other. In each field they determine what a scientist in the field would consider a legitimate scientific statement. Orthogonal to this distinction, it is important to differentiate between (a) underdetermination by all possible empirical data and (b) underdetermination by the currently available data.

The concepts of underdetermination which have played the most substantial roles in the philosophy of science up to now were exemplifications of versions (1b) and (2a). Hume's claim that a given data set does not logically imply any future events refers to the available evidence and asks for logical possibilities. Thereby, it constitutes the classic example of a claim of type (1b) underdetermination. The

---

[3] Using the name "scientific underdetermination" is not meant to suggest that other types of underdetermination are of no or lesser interest to a philosophical analysis of the scientific process. The name has been chosen based on the point made above that scientific underdetermination is the ̱ind of underdetermination most relevant to the acting scientist herself.

Quine–Duhem thesis, which holds that any individual statement can be made compatible with any new empirical data by making appropriate changes within the wider conceptual framework, would be another example for a claim of (1b) underdetermination. Quine's statements regarding the underdetermination of scientific theory building in Quine (1970, 1975) deal with the options for constructing scientifically viable and empirically equivalent theories. Therefore, they exemplify (2a). A wide range of recent philosophical thoughts about underdetermination also falls into this category. Laudan and Leplin (1991), van Fraassen (1980) or Sklar (2000) would be prominent examples.

Version (2b) arguably is the one most relevant to the scientist who searches for new theories. In order to be able to develop theoretical schemes at all, the scientist must take for granted the validity of a basic principle of induction, the existence of a coherent theoretical scheme capable of describing the phenomena in question and some vague assumptions about the universality, predictive power and lack of ad-hoc-ness of that scientific scheme. She thus must take for granted a framework of ampliative rules of the scientific method. Science thus is concerned with underdetermination of type (2). Moreover, the scientist aims at producing successful predictions based on the currently available data. The scientific search for new theories must be based on the empirical evidence currently available, which implies that the kind of underdetermination of interest to the scientist in this context is of type (b). The claim of (2b) under-determination in a certain field at a given time asserts that it would be possible to build several or many distinct theories which qualify as scientific and fit the empirical data available in that field at the given time. Since these alternative theories are merely required to coincide with respect to the presently available data, they may well offer different predictions of future empirical data which can be tested by future experiments. It is type (2b) underdetermination due to the existence of such empirically distinguishable theories which will be of primary interest in the following analysis.

Type (2b) underdetermination so far has played a less prominent role in the philosophy of science than (1b) or (2a). It is by no means unknown in the philosophical literature, however. Lawrence Sklar (1975) may have been the first to discuss it as a distinct form of underdetermination under the name transient underdetermination. Kyle Stanford (2001, 2006) has recently emphasized the importance of transient underdetermination for the scientific realism debate. The reason why the term "transient underdetermination" is not adopted in this book has to do with the profound differences between the perspective on underdetermination chosen by Sklar and Stanford and the one to be developed in this book. Sklar and Stanford stay within the canonical paradigm of theory assessment and understand underdetermination as something that must established by actually finding alternative theories. Once alternative theories have been found, underdetermination with respect to those theories can be removed by carrying out empirical tests which decide between these alternative theories. Seen from this perspective, the use of the term transient underdetermination for type (2b) underdetermination appears natural. The core claim of this book, to the contrary, will be that the degree of type (2b) underdetermination can be assessed without knowing the alternative theories. Therefore it becomes important to understand the degree of underdetermination in terms of the number of *possible* alternatives, irrespective of the question whether those alternatives are known or not. "Transient underdetermination" from this perspective would suggest that all *possible* alternative theories of today can be removed by experiment in the future so that, at some stage, there will remain no alternative to the surviving scientific theory any more at all. But this is obviously not what Sklar, Stanford or I want to assert by claiming type (2b) underdetermination.[4] Therefore, using the term "transient underdetermination" for type (2b) underdetermination would be highly misleading in the context of this book, which is why I use the term "scientific underdetermination" instead.

The distinction introduced in Section 2.1 between scientific speculations and scientific knowledge can now be motivated by referring to the principle of scientific underdetermination. Well-established scientific theories are those whose distinctive predictions[5] have been experimentally well tested and confirmed in a certain regime. The general viability of the theory's predictions in that regime is considered a matter of inductive inference.[6] Speculative theories, on the other hand, are those whose distinctive predictions have not yet been experimentally confirmed. Even if a speculative theory fits the currently available experimental data, its distinctive predictions might well be false due to the scientific underdetermination principle.

Scientists and philosophers of science endorse the principle of scientific underdetermination for a number of reasons. First, scientific underdetermination is supported by the vaguely instrumentalist spirit inherent in the prevalent understanding of the scientific process. According to this understanding, the scientist builds theoretical structures which reflect the regularities observed in nature up to some precision and tunes the involved free parameters in order to fit the quantitative details of observation. The successful construction of a suitable theory for a

---

[4] In fact, the question whether or not all possible alternatives can be excluded so that only one theory remains will be the main topic of Part III of this book.

[5] That is, predictions, whose experimental confirmation would be direct empirical support for the novel theoretical claims of the theory.

[6] A scientist's formulation of the notion of well-established theories can be found e.g. in Weinberg (2001).

significant and repeatedly observable regularity that characterizes the world is assumed to be just a matter of the scientist's creativity and diligence. If it is always possible to find one suitable scientific theory however, it seems natural to assume that there can be others as well. Different choices of theoretical structure must be expected to exist which have coinciding empirical implications up to some precision in the observed regime if their respective free parameters are fixed accordingly. The principle of underdetermination follows from this.

Furthermore, the assumption of scientific underdetermination constitutes a pivotal element of the modern conception of scientific progress. If science proceeds, as emphasized e.g. by Kuhn (1962) or Laudan (1981), via a succession of conceptually different theories, all future theories in that sequence must be alternative theories which fit the present data and therefore exemplify scientific underdetermination.[7] Theoretical progress without scientific underdetermination, to the contrary, would have to be entirely cumulative.[8]

In addition, there are many concrete examples of scientific contexts where the available empirical data is known to allow a wide spectrum of conceptualizations with substantially different empirical predictions with respect to new data. In particle physics, to give one example, we know that many different extensions of the particle physics standard model are compatible with the presently available data.

The importance of the principle of scientific underdetermination thus is well established. Scientific underdetermination cannot be taken to be entirely unconstrained, however. If it were, that is if all imaginable regularity patterns of empirical data could be fitted by fully satisfactory scientific theories, no correct predictions of new phenomena could ever be expected to occur. One would rather be led towards understanding the specific predictions offered by the presently available theory as one accidental "pick" among an infinite number of theoretically viable options. In this light, it would seem entirely unreasonable to take those predictions seriously. It is a fact, however, that successful predictions of new phenomena do happen frequently in advanced science. Scientific underdetermination thus must be assumed to be limited in some way.

---

[7] It should be emphasized that the last statement would lack distinctive meaning if based on the most radical reading of Kuhn's incommensurability thesis. The statement relies on the assumption that adherents of the successive theories referred to can find a consensus with respect to the scientific characterization of the collected empirical data. In the context of particle physics, which shall be analyzed in the present work, this assumption clearly seems justified as adherents of all existing particle physical theories share the same understanding of the implications of specific particle experiments for theory building.

[8] The scientific underdetermination principle is closely related to the pessimistic meta-induction of Laudan (1981) but does not share the latter's anti-realist claims. It reflects an ontologically neutral assessment of the status of scientific theories that is fairly uncontroversial in recent science and philosophy of science.

Does science have to care about the nature and extent of those limitations? Is the assessment of those limitations part of the scientific process? Does it constitute a theoretical element of scientific theory assessment? As we have argued above, the classical empirical paradigm of theory assessment does not allow for any such role. It will be the task of the present book to argue otherwise.

3

The assessment of scientific underdetermination in string theory

3.1 Connections between theory assessment and scientific underdetermination

It may seem surprising at first glance that scientific underdetermination should play any role at all in assessing string theory. Section 2.3 introduced scientific underdetermination as underdetermination of scientific theory building by the available empirical data. String theory, however, has remained entirely unconfirmed by empirical data up to now. The question of underdetermination would arise more naturally if we were dealing with a theory that has found some empirical confirmation and would ask the question how many alternatives could be constructed in agreement with the available empirical data.

In order to understand the rationale of the argument in the given case, one must remember that string theory has been developed based on and informed by empirical data. It has been constructed in order to provide a universal theory that underlies, and has as its approximations in certain regimes, two theories (general relativity and gauge field theory) which have been empirically confirmed. At present, the data that confirms those theories does not constitute empirical evidence for string theory because it cannot be predicted based on the current understanding of string theory. However, since string theory conceptually relies on the given empirical data, the conditions for applying assessments of scientific underdetermination are fulfilled. A complex web of posits, conjectures and physical analysis lies between the available empirical data and the theoretical concept of string theory. As discussed in Section 1.2, it is the threat of scientific underdetermination that normally prevents trust in theoretical speculations whose connections to available empirical data are that distant. In this light, it is natural to expect that only an assessment of scientific underdetermination can increase the trust in such theories. Claims of limitations to scientific under-determination have the potential of strengthening the inferential connection between the empirical data that has motivated the theory's construction and the theory itself. In the following we will argue that the three arguments for the viability of string theory presented in Chapter 1 can be understood in this sense.

Let us first have another look at NAA, the no alternatives argument. The observation that no equally satisfying solutions to a given scientific problem have been discovered by scientists even after a long and careful search for alternatives can be interpreted in two different ways. First, one could conclude that the depth and scope of the scientific analysis at the time just have not been sufficient for finding the appropriate alternatives. Some alternatives within the chosen framework may have been overlooked or some of the foundational postulates which have defined the framework for the overall search may require modifications in order to allow alternative solutions. That conclusion would offer no reason for believing that the lack of known alternatives to the available theory has any implications for that theory's chances of being empirically viable. Limitations to human intellectual capacities provide no good argument for the viability of the theory scientists came up with. The situation may also be interpreted in a different way, however: one might conjecture a connection between the spectrum of theories the scientists came up with and the spectrum of all possible scientific theories that fit the available data. The observer who chooses this second path takes the fact that scientists have problems finding alternatives as a sign that not too many alternatives are possible in principle. In other words, she concludes that scientific underdetermination is significantly limited. Based on the additional assumption that the phenomena in question can be characterized by a coherent scientific theory at all, the conjecture of significant limitations to scientific underdetermination can then enhance the trust in the available theory's viability: if a viable scientific theory exists and only very few scientific theories can be built in agreement with the available data, the chances are good that the theory actually developed by scientists is viable. The step from an observation about the present human perspective to a conclusion regarding the overall spectrum of possible scientific thinking is by no means trivial and raises deep philosophical questions. Still, it is a necessary precondition for using the argument of no choice as an argument for a theory's viability.

UEA, the argument of unexpected explanatory coherence, is related to an assumption of limitations to underdetermination as well, albeit in a less direct way. The initial observation in the given case does not characterize the human activity of finding and developing alternative theories like in the case of the no alternatives argument. It rather deals with properties of the theory itself. The reasoning relies on the observation that some explanatory connections provided by the theory were not aimed at during theory construction but have emerged after closer analysis of the theory's structure. The argument thereby mirrors the canonical reasoning for a theory's viability based on novel empirical confirmation. The importance of distinguishing between unexpected explanations and intended ones can be argued for in analogy with the distinction between novel data and data that has entered the process of theory construction. If physicists were searching for a microscopic explanation of black hole entropy and came up with a theory whose only merit was to offer such an explanation, it would not make sense to take that merit as a strong indication for the theory's validity. Since that was what people were searching for, no one could be surprised that they eventually came up with a theory of that kind. If, like in the case of string physics, the theory was developed for other reasons, however, the gratuitously received extra merit is of significance. It gives the impression that physicists are on the right track.

But why is it that surprising explanatory power can reasonably be taken to enhance a theory's chances of being viable? Naturally, a theory that solves a higher number of conceptual problems of predecessor theories can be considered more likely to be viable. This does not account for the additional aspect, though, that the theory was not constructed to provide those solutions. In order to understand the extra value of surprising explanatory power, let us first imagine a scientist who approaches the problem (p1) how to unify gravity and gauge field theory under the assumption that solutions to that problem are abundant. Let us further assume that she takes solutions to other seemingly independent problems like the problem (p2) how to explain the scale hierarchy between electroweak scale and Planck scale or the problem (p3) how to acquire a microscopic understanding of black hole entropy to be abundant as well. Finding a theory that solves one of those problems from her perspective does not imply that this theory is physically viable. The physically viable solutions (i.e. the ones whose empirical core predictions are correct) may well be among the many other possible theories. Now let us assume that this physicist finds string theory to be a solution to the unification problem (p1). Without additional information, the described physicist does not have any reason to expect that string theory will also be a solution for problems (p2) and (p3). If the physicist adheres to a general form of scientific optimism and assumes that there is an empirically fully adequate scientific theory that covers the research field to which all three problems belong, she must expect that among the large number of solutions to each of the problems there will be at least one which is empirically fully adequate and therefore solves all three problems at once. There may even be a number of other theories which solve all three problems as well. It is to be expected, however, that a vast number of theories, probably most theories which offer a solution to any of the three problems at all, will only solve one of the problems. The given physicist has no reason to believe in advance that string theory is not an element of the latter group.

Now compare this with the perspective of another string theorist who is a strict supporter of NAA and believes that there are no possible alternatives to string theory. Based on the principle of scientific optimism introduced above, string theory thus must be taken to be the true theory. Since the given string physicist must expect that the true theory solves all problems, she is forced to expect that string theory will solve all problems. In other words, the expectation of explanatory interconnections the theory was not constructed to provide is natural for someone who believes in the most rigid limitations to scientific underdetermination while it is unnatural for someone who believes in the unconstrained abundance of theoretical solutions. Finding unexpected explanatory power therefore supports the conjecture of limitations to underdetermination. UEA can strengthen the case for NAA.

Even in conjunction, however, the two arguments NAA and UEA are not conclusive. As argued in Chapter 1, unexpected explanatory power could have other reasons than a lack of alternatives to the corresponding theory. It could arise due to so far insufficiently understood theoretical interconnections at a more fundamental level, of which the theory in question is just one exemplification among many others. If so, the observed unexpected explanatory power would indicate the viability of that underlying more fundamental principle rather than the viability of the theory itself. In order to distinguish between the two cases, it would be helpful to get a better grasp of the actual chances of empirical success of theories which show a strong pattern of unexpected explanatory success.

MIA, the meta-inductive argument from the success of other theories in the research program, provides information to that end in an intricate way. As discussed above, MIA is an empirical argument. The empirical data deployed does not serve as an empirical test of the theory under consideration, however. It is used at a meta-level, providing an empirical test of the strategies on non-empirical theory assessment. If those strategies are found to be regularly successful, they become more trustworthy. The empirical observations that provide the basis for MIA thereby increase the trust in so far empirically unconfirmed scientific theories which are supported by the given strategies.

It is of crucial importance that MIA conveys a message of limitations to underdetermination already on its own terms. The line of reasoning to that end is similar to the one presented above in the context of UEA and has already been briefly addressed at the end of Chapter 2. Let us, in analogy to the discussion of argument UEA, consider an observer who assumes that there are many different possible theories which can account for the empirical data available at some point. Now, let us look at predictions of novel phenomena offered by those theories. (In the context of particle physics, such predictions will primarily be concerned with new elementary particles and their properties.) Without further assumptions, the given observer must expect that the various theories will offer many different predictions with regard to the next generation of empirical tests. On these grounds, chances seem rather small that the theory developed by scientists in order to account for the available data will offer correct predictions with respect to the upcoming empirical tests. The assumption of limitations to underdetermination, on the other hand, can provide an explanation of predictive success: the fewer the scientist's options for constructing scientific theories that fit the available data and offer different predictions regarding the upcoming experimental tests, the better are her chances of selecting one of those theories which give correct predictions of the results of those tests.

The question remains whether predictive success could be explained in other ways as well which do not rely on claims of limitations to scientific under-determination. Some potential alternative explanations come to mind. For example, one might assume that specific properties like simplicity or beauty guide scientists towards finding predictively successful theories. If scientists were indeed striving for simple and beautiful theories and if there were a deep connection between simplicity or beauty and predictive success, that connection might serve as an explanation for the predictive success of the theories scientists develop.

Would this constitute a viable alternative explanation of predictive success? One may have the general objection that the suggested explanation suffers from the notorious problems to define simplicity and beauty in the given context. It is by no means self-evident that the theory development in fundamental physics or any other field moves towards "simpler" theories in any sense that goes beyond higher universality. If no development towards simplicity in a significant and independent sense can be confirmed, however, it would seem untenable that higher simplicity is correlated with higher predictive accuracy. In the context of fundamental physics, it is not clear either, whether it makes sense to attribute to the scientists the conceptual freedom to choose between simpler and less simple or beautiful and less beautiful theories. More often than not, scientists must be content with finding any coherent solution to their conceptual problems at all.

Even if we assumed that definitions of simplicity and beauty can be provided in a meaningful way and do play a significant role in theory building, however, closer inspection reveals that the corresponding explanation of predictive success does not constitute an *alternative* to limitations to scientific underdetermination but must rely on the assumption of limitations to scientific underdetermination itself.

Let us assume that scientists have found a theory with a given degree of simplicity and beauty. Explaining the predictive success of that theory on that basis must rely on the assumption that only few alternative theories with similar or higher degrees of simplicity or beauty exist. A certain degree of beauty or simplicity thus is turned into a condition which has to be fulfilled by those theories with regard to which limitations to underdetermination are assessed. In other words, invoking simplicity, beauty or similar criteria which are expected to be fulfilled by predictively successful theories merely changes the framework of conditions within which assessments of scientific underdetermination are carried out. An explanation of predictive success that works without any reference to limitations to scientific underdetermination would have to abstain from imposing any such criteria and assume a human capacity to find predictively successful theories by means which transcend the analysis of the available empirical data and the structural properties of potential theories. It would thus amount to positing a kind of magical "truth detector" that finds the true theory just because it is true. This, however, seems alien to scientific reasoning. Whenever scientists do manage to develop various conceptually satisfactory solutions to a given scientific problem, all those theories must be considered as viable candidates for the valid solution of that problem. No magical "truth detectors" can be deployed in order to find the true solution.

Assuming limitations to scientific underdetermination therefore indeed looks like the only satisfactory explanation of the fact that scientists regularly find theories which are predictively successful. Inference to the best explanation then can lead from the observation of regular predictive success in a scientific field towards the conjecture that scientific underdetermination is limited in that field. Any instance where empirical predictions get confirmed by empirical tests thus can be understood as an indication of limitations to underdetermination in the given regime.

The next step of MIA consists in making the meta-inductive inference that regular predictive success in a research field justifies the assumption that future predictions of a similar kind will be correct as well. To be applicable, the inference must rely on a reasonable understanding as to what can count as a prediction of a similar kind. Obviously, regular predictive success in some research field does not imply that every theory that is constructed in the field in agreement with the known data is likely to be predictively successful. Any inference from one theory's predictive success to the probability of another theory's empirical viability must be based on reasons to believe that the new theory is in a significant way comparable to the earlier ones. This is where limitations to underdetermination become a crucial bond between the three arguments of non-empirical theory assessment. We have noted that predictive success can be linked to limitations to scientific underdetermination. We have further noted that NAA and UEA offer independent reasons for believing in limitations to underdetermination in specific contexts. Above, we have seen that MIA can be understood in terms of limitations to scientific underdetermination as well. Therefore, we have good reasons for taking the conditions corresponding to NAA and UEA to be good criteria for selecting the group of theories within which we can make legitimate inferences based on MIA. If novel predictive success occurs regularly when theories fulfil criteria for NAA and UEA, it is justified to expect with some confidence that it will occur with respect to another theory that satisfies conditions for NAA and UEA as well.

The interrelated web of reasoning presented above consists of arguments that mutually support and strengthen each other and provide a machinery of theory assessment that is based on empirical data. NAA and UEA provide criteria for the inference carried out in MIA. MIA, on the other hand, may be understood in terms of an empirical test of the criteria defined by arguments NAA and UEA. If theories which were (or could have been) considered probably viable based on NAA and UEA turn out to be empirically successful in many instances in a scientific field, that can be taken as empirical corroboration of the viability of the criteria provided by NAA and UEA.

The question now arises: can this kind of reasoning be acknowledged as a critical scientific method which has some legitimacy for providing a foundation for theory assessment? Any scientific strategy of theory evaluation must specify the circumstances under which its criteria of scientific success speak against a theory's viability. Moreover, there must be specifiable possible circumstances under which the status of the strategy itself as a viable strategy of theory assessment weakens on its own grounds. In the given case, it would clearly be insufficient to specify empirical data that refutes the theory under scrutiny. Since the entire enterprise has been started in order to allow for trustworthy theory assessment in the absence of empirical data predicted by the theory in question, the scenarios under which the assessment speaks against the theory's viability must be of a non-empirical nature as well.

Indeed, it is possible to list a number of scenarios which would significantly reduce or even nullify the trust in string theory based on purely theoretical evaluation criteria. One can also find scenarios in which the strategies of non-empirical theory assessment would lose much of their appeal. Let us look at such scenarios one by one.

NAA relies on the observation that no conceptually equally satisfactory alternatives to string theory are found. It would have to be withdrawn whenever a conceptually satisfactory alternative to string theory was in fact discovered. Any such alternative would thus significantly reduce the trust in string theory

UEA would lose strength if some of the interconnections in question turned out to be based on deeper and more general patterns of reasoning which were not univocally related to string theory. Let us assume that string theory provides an explanation for some structural characteristics we find in physics which did not have a satisfactory explanation before. It may happen that, after a while, we understand that this explanation is based on a deeper physical principle X that could be extracted without any reference to string physics. String theory in that case just would have been the context where we first understood a principle X that was in fact far more general than the specific theory. In that case, we must not be surprised that string theory provided the given explanation since, due to X, any alternative theory would have provided that explanation as well. Therefore, we could not use the fact that string theory provides the explanation in question as an argument for the viability of string theory but only for the viability of more general assumptions which generate principle X. In fact, considerations along these lines are being discussed in string physics and contribute to a better understanding of the significance of the argument of unexpected explanatory interconnections. For example, it has been argued that microscopic calculations of black hole entropy can be calculated based on general principles without using a string theoretical framework (see Strominger, 1998, and Carlip, 2008).

MIA can lose power based on new empirical data regarding other theories. As argued already in Section 1.3, trust in string theory would be substantially reduced if theories which seemed to have no alternatives turned out to make false predictions. If no Higgs particle had been found during the LHC experiments at CERN, many observers would have raised the question how confident one could be regarding the remote claims of string theory if one could not even correctly predict the existence of the Higgs particle.

On a more general basis, theoretical arguments in favor of string theory could also be weakened by other developments which would cast doubt on the chances of success of the research program. For example, an improved theoretical understanding of string physics might reveal theoretical weaknesses which change the theoretical assessment of the theory's chances of being fully consistent and thus physically viable. Furthermore, an interruption of theoretical progress over a long period of time could raise doubts whether a more complete theoretical understanding of string physics were attainable at all.

New information can also reduce the trust in the method of non-empirical theory assessment itself. Generally, each instance where a theory has been taken to be probably viable based on non-empirical theory assessment and then lost its trustworthiness again due to new evidence, be that evidence empirical or non-empirical, weakens the status of non-empirical theory assessment in the very same way that it is strengthened by instances which show its consistency or predictive power.

To conclude, we face a situation where theories can be supported as well as weakened based on non-empirical evidence. The same kinds of analysis which can provide theory confirmation without finding data that reproduces the theory's predictions, can, under different circumstances, amount to the disconfirmation of the theory without having found data that contradicts the theory's predictions.[1] Therefore, it seems indeed possible to grant non-empirical theory assessment the status of viable scientific reasoning.

The skeptic regarding the significance of non-empirical theory assessment might still have one fundamental objection. Nothing of what has been said so far, she might say, has dissolved the fundamental difference between the empirical confirmation of a scientific theory and non-empirical theory assessment. Even if one may find reasons for giving some significance to the latter, the fundamental epistemic difference between observing a phenomenon on the one hand and inferring a statement's validity based on circumstantial evidence on the other must be upheld. As a matter of principle, trust in an empirically unconfirmed theory can never be compared to trust in a theory that has been empirically confirmed.

The answer to this objection is twofold. First, the presented arguments do not suggest that non-empirical theory assessment can replace empirical confirmation or assume the very same status within the structure of scientific reasoning. Since non-empirical theory assessment crucially relies on empirical confirmation within the research field, it is in an important sense secondary to the latter.

Second, however, the difference in epistemic status between empirical confirmation and non-empirical theory assessment is less fundamental than one might assume at first glance. The substantially stronger position of empirically confirmed theories compared to theories supported by non-empirical reasoning is not an immediate consequence of the scientific method but rather a non-trivial empirical fact about the world we face.

We often trust empirically well-confirmed theories to the extent that we are willing to bet our lives on their viability: we use aeroplanes or bridges based on our understanding that they have been built according to well-confirmed scientific theories. In fact, we would call irrational anyone who refused to do so. A person that refuses to step on a bridge because of doubts regarding the involved physical laws would be considered profoundly strange.

To the contrary, few people would be ready to trust a theory to the same extent based on non-empirical theory assessment. But let us imagine, for a moment,

---

[1] In fact, the latter scenario is quite uncontroversial in some cases; finding out that a theory constitutes just one out of a million theoretical options to fit the available data makes its viability quite nearly as unlikely as finding data that contradicts its predictions.

some "strange world" that is very different from the world we live in. Physicists in this "strange world" have developed theories for 10 000 years and every single theory they have developed in a mathematically coherent way without being able to find an alternative eventually has turned out to be empirically viable. In that world, NAA (the argument that the observation that scientists have not found an alternative to a theory indicates a theory's viability) would be considered every bit as trustworthy as standard inductive inference based on empirically well-confirmed scientific theories. It would be obvious in this "strange world" that physicists have the capacity to exhaustively consider possible alternatives even if no one had come up with an explanation how they do it. A person that doubted a theory that was confirmed by NAA would seem quite as weird in the "strange world" as a person that does not believe in the stability of bridges does in ours.

Obviously, we do not live in a "strange world." In our world, physicists do have problems finding existing alternative theories and the confirmation value of non-empirical evidence is more precarious and difficult to assess – which is why the debate about the significance of non-empirical theory assessment is highly non-trivial and by no means stupid. But the mechanisms which can lead us towards acknowledging the confirmation value of non-empirical evidence remain the same in principle in our world as in the "strange world." If science enters a phase where such arguments work better than in earlier periods or other contexts of scientific reasoning, it is therefore rational and necessary to modify the status of non-empirical theory assessment accordingly. Non-empirical theory assessment in our world obviously is more cumbersome, less stable and less precise than straightforward theory assessment based on empirical testing. Assessing the degree to which it is, however, constitutes a crucial scientific task.

3.2 The framework for claims of limitations to scientific underdetermination

Any statement asserting that only few scientific theories can be developed which fit a given data set must be based on a definition of what counts as a scientific theory. Without specifying a sufficiently rigorous framework within which we count possible theories, we cannot meaningfully claim that the number of those theories is limited. If one would not take for granted a commitment to the validity of inductive reasoning, to take the most obvious example, theory building would be free to predict any pattern of future events imaginable and thus could not be taken to be limited in any meaningful sense. In Chapter 2, we established a framework for statements on underdetermination by restricting our analysis to ampliative scientific reasoning. The term "ampliative" denoted the assumption of a certain set of conditions which have to be fulfilled by a theory in order to be called scientific. Statements on limitations to scientific underdetermination must be understood within this framework of presumptions. Two important questions arise here. What is the basis for restricting the considerations on limitations to underdetermination to the framework of ampliative scientific reasoning? And to what extent is it necessary to specify the conditions of ampliative reasoning in order to provide a solid foundation for statements on scientific underdetermination?

In order to get a clearer understanding of those questions, let us for the moment forget about the ampliative conditions of scientific reasoning mentioned above and start with the most basic statement. A claim of limitations to underdetermination must be made within some framework that defines which kinds of theories are addressed by the claim. The selection of an adequate framework then must satisfy two conditions which drag in opposite directions. On the one hand, it must be sufficiently strong for allowing significant statements on limitations to underdetermination. On the other hand, it must be sufficiently wide for allowing high confidence that a viable scientific description of the phenomenology to be predicted can be found within its limits.

If the second condition were not met, we would have a nice statement on limitations to underdetermination but could not use it for the purpose of establishing that a theory has good chances of being viable. In order to understand the problem, let us imagine that we choose a specific set of general physical principles – for example the principles of gauge field theory in particle physics – as a framework for a statement on limitations to underdetermination. Since we understand the gauge principle fairly well, this would be a convenient choice for making strong statements on limitations to underdetermination. However, the question would arise as to how probable it is that the next generation of empirical data can in fact be described successfully by a gauge field theory. This new question would have to be assessed by starting a new level of assessment of underdetermination: we would have to assess the number of scientific theories which are not gauge theories, reproduce the available data and give predictions regarding new data which differ from those of the gauge theories. If there are many of them, we would have to consider it likely that the viable theory will not be a gauge theory. After all, we have no good reason to believe that the scientific principles we have happened to develop at this point are more likely viable than other principles which offer equally satisfactory descriptions of the available data. In order to carry out that second level assessment, however, we would have to introduce a second level of scientific principles that provides the framework for that assessment. Those more general scientific principles may be questioned again and we face a threat of an infinite regress. Using known scientific concepts as frameworks for statements on limitations to underdetermination thus does not help in establishing any assessment of a theory's viability.

The only way to avoid this problem consists in retreating to a framework of presumptions that is general enough for justifying trust in its capacity of allowing viable descriptions without further assessments of underdetermination at a meta-level. Using ampliative criteria of scientific reasoning looks like a natural choice since it conflates trust in the framework's legitimacy with trust in the success of scientific reasoning. The scientist thus can say that, qua being a scientist, she believes in the viability of the scientific method and takes it for granted. It is a presumption closely related to the belief in the general viability of an inductive principle: if a strong inductive inference that relies on our present body of scientific knowledge fails, the scientist takes it for granted that this can be explained based on a new scientific theory.

The problem may look less threatening now, but it has not vanished. The question remains as to how stable the present criteria of scientific reasoning must be. In fact, we know of a number of criteria which would have been taken to be essential elements of any legitimate scientific theory by most scientists at an earlier point in history but have been abandoned today. An example in case is the principle of determinism that was toppled by quantum mechanics. In this light, tomorrow's scientific theories may well lie beyond today's framework of ampliative criteria of scientific reasoning. The scientist nevertheless has a method of retaining a reasonably stable framework. She can assume, based on extensive and coherent historical data, that most of the principles of scientific reasoning valid today will survive the next stages of empirical testing. She thus can refer to a vague notion of a core of today's scientific principles which will remain viable at the next stages of scientific enquiry. That core of present-day scientific principles then can be deployed as a framework for the claim of limitations to scientific underdetermination.

It is important to understand that a precise specification of scientificality conditions is not necessary for carrying out an inference to limitations to underdetermination in a coherent way. The inference is based on the observation of frequent predictive success in the research field. On that basis, it is inferred first that scientists apply a framework of scientificality conditions which are quite stable, and second, that scientific underdetermination is limited within that framework. The precise nature of the framework is never made explicit in the argument.

One further consideration may alleviate doubts about basing scientific reasoning on foundations of such a shaky nature. The deepest foundations of scientific reasoning always tend to be the ones which are the most difficult to stabilize philosophically. Just think of the problem of induction, the intractability of which stands in stark contrast to its pivotal role in all scientific reasoning. Part II of the book will sharpen the focus of this comment. It will turn out that assessments of limitations to underdetermination are by no means confined to the context of empirically unconfirmed theories. Rejecting its capability of generating scientific knowledge due to its unstable philosophical foundations would threaten much of what the scientific observer is used to take as stable scientific knowledge.

3.3 The scope of the three arguments

The three arguments of limitations to scientific underdetermination, if applicable, suggest a scarcity of possible alternatives to a given empirically unconfirmed theory that has been developed for theoretical reasons. Based on the three arguments, limitations to underdetermination are taken to be sufficiently strong for justifying the belief that the theory in question will get empirically confirmed once the critical experimental tests can be carried out. How can we understand the scope of that claim?

High energy physics knows one crucial parameter that defines the range of experimental testing: the energy provided for generating new particles in deep inelastic scattering. If a high energy theory predicts new particles at some characteristic energy scale, predictive success at that scale can be explained by a scarcity of theoretical alternatives which are predictively distinct at that scale. Meta-inductive reasoning then can lead from the observation of regular success of that kind to scarcity of alternatives with regard to the next step of empirical testing at some higher energy scale.

Note that this inference does not amount to an absolute limit to the number of empirically distinct scientific theories which fit the available data. This point can be seen most clearly in MIA, where empirical testing at a meta-level can only provide support for limitations to underdetermination regarding the immediate tests of core predictions made by the theory in question. The fact that those core predictions have been confirmed does not justify the claim that the theory will never be superseded. Attributing a low probability to the existence of alternatives which are predictively different at the next stage of empirical testing is fully consistent with the expectation that an infinite sequence of ever higher energy scales lies beyond that next empirical step. That, however, would imply that any number of alternatives could be made probable by taking into consideration a sufficiently high number of future experimental steps. The empirical data establishes only one restriction to those new theories: they must have the known theory as their low energy effective theory (i.e. as a good approximation up to some level of accuracy) at that theory's characteristic energy scale.

Let me illustrate this point by looking at the prime example of predictive success in the high energy physics research program, the standard model of particle physics. As discussed in the context of MIA, the predictive success of the standard model is constitutive for the current trust in empirically uncon-firmed theories like string theory. However, the success of the standard model did not exclude (and its creators did not intend to exclude) that more funda-mental modifications of the present physical theories might also be capable of curing the problem of the renormalizability of nuclear interactions.[2] For exam-ple, instead of relying on gauge symmetries, one could have ventured already in the 1960s to make the step towards extended elementary particles, a step that was later realized by string theory. To believe in the standard model in the early 1970s merely meant assuming that any more far-reaching change of physical postulates, in as much as it would be successful, would itself imply the standard model predictions. This assumption has been vindicated by the subsequent development of physics. Extended elementary particles have emerged as a (potential) next scientific step but it turned out that their introduction, if con-sistently done, implies gauge theory as well.[3]

Relying on this kind of example, the physicist who applies MIA can only infer the new theory's viability at the next steps of empirical testing (irrespective of the chances for actually taking those steps in the foreseeable future). MIA does not address the question as to whether the theory to be assessed is absolutely true, whether it is empirically adequate under all possible evidence or the like. It does not imply that the theory will never have to be improved or extended by new theoretical principles or even be superseded by a fundamentally different theo-retical concept. Formulated in terms of scientific underdetermination, MIA does not support the claim of an absolute limitation of the number of possible scientific theories. It merely establishes the claim that the number of theories which can be distinguished at the next steps of empirical testing is probably strongly limited. Theories which cannot be distinguished from each other at the next steps of empirical testing are not counted as different theories. We want to call a claim that asserts a limited number of possible theories which fit the available data and are

---

[2] In fact, in the case of the standard model technical reasons did suggest early on that the theory at some stage would have to be embedded within a wider theoretical framework. Those considerations were instrumental for developing those theories beyond the standard model which were presented in Chapter 1.

[3] String theory can only be consistently formulated in a way that makes its low energy effective theory a gauge theory (see e.g. Polchinski, 1998, Chapter 12).

empirically distinct at the next stages of empirical testing a claim of *local* limitations to scientific underdetermination.

UEA in itself does not reach out beyond the limits just described either. Unexpected explanatory interconnections pertaining to the given theory's characteristic scale may be explained by limitations to scientific underdetermination at that scale but do not exclude the possibility that new theoretical options which share the same explanatory potential might open up at higher scales.

The situation regarding NAA is a little more complex. In its most radical interpretation, NAA can be interpreted as suggesting that no experimentally viable alternative to the theory in question exists at all. In other words, NAA might be taken to suggest that the theory in question will never be superseded by any new theoretical framework because no alternative scientific theory that can reproduce the available data is possible. Such a radical claim shall henceforth be called a final theory claim. In most contexts of physical research, a final theory claim would be rendered highly implausible by the fact that the theory under investigation is not totally universal. A more advanced theory that constitutes a next theoretical step towards higher universality, however, would constitute a conceptual alternative of the kind ruled out by the final theory claim. For all theories which are not totally universal, the more modest interpretation in terms of *local* limitations to scientific underdetermination thus seems appropriate.

Now, string theory is a universal theory of all interactions. In this light, it does not seem absurd in the case of string theory to make a final theory claim. However, as argued above, the empirical corroboration of NAA must be based on MIA, which in turn can only establish local limitations to scientific underdetermination. Therefore, the significance of NAA as an argument for a final theory remains entirely unsupported within the framework discussed up to now. In order to make a serious case for a final theory, one would need other independent arguments which make plausible the additional step. We will pick up this thread of reasoning in Part III of this book where we will present two kinds of arguments which actually hint in this direction.

3.4 Non-empirical theory assessment as inference to the best explanation

We have now developed the full line of reasoning that infers a theory's chances of being viable at upcoming stages of empirical testing from assessments of scientific underdetermination. In the following two sections, we are going to relate that argument to two influential schematizations of scientific reasoning in current philosophy of science. In this section, we will define local assessments of scientific underdetermination in terms of inference to the best explanation (IBE). The following section will analyze it from a Bayesian perspective.

The three arguments of non-empirical theory assessment which were discussed in this chapter are all examples of IBE. IBE is a form of inference to the truth or viability of statements that was first defined by Peirce and is generally taken to be constitutive of scientific reasoning. It goes beyond deductive reasoning and enumerative induction by inferring the viability or truth of statements based on the claim that they constitute the best explanation of a certain set of empirical phenomena. For the sake of simplicity we will henceforth discuss the subject in terms of truth, keeping in mind that IBE-type reasoning may also be deployed with regard to more limited scientific goals than truth. Peter Lipton (2004) and Alexander Bird (2007), who have offered recent analyses of IBE, distinguish two phases of inference to the best explanation. A first phase consists in finding a number of potential explanations from the pool of all possible explanations. A second phase then ranks the explanations found in phase one and eventually, if certain conditions are met, selects the "best" one and infers its truth.

The success of IBE obviously requires that the scientists select a set of explanations in phase one that includes the true explanation. In order to trust IBE, scientists must be confident that the true theory does not belong to the set of unconceived alternatives. Strictly speaking, IBE thus always involves an element of reasoning that amounts to assessments of scientific underdetermination. Normally, however, reconstructions of IBE do not focus on that aspect but are much more concerned with the selection process of the best theory among the selected alternatives (see Lipton, 2004, and Bird, 2007). The implicit idea seems to be that, if the best explanation is very convincing indeed, we may be justified to disregard the threat of unconceived alternatives. As an immediate consequence of this perspective, IBE is taken to be convincing only in cases where the best-known explanation is known to be very good. In particular, the best-known explanation should offer convincing explanations for all or most phenomena in the relevant domain and all or most of its implications should be known to cohere with the data. The quality of an explanation whose core predictions have not been confirmed yet cannot be assessed to a sufficient degree, which means that theories lacking empirical confirmation normally are not taken to merit trust based on IBE. IBE as normally understood thus is in agreement with the canonical understanding of theory assessment.

The present analysis is concerned with situations where theories are trusted despite a lack of confirming data. The general scheme of IBE is fully applicable in that case. However, the reasons for trust in IBE cannot be derived from the quality of the known scientific explanation alone. It must involve assessments

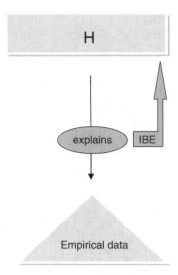

Figure 3.1 IBE at the level of scientific analysis.

of scientific underdetermination of the form discussed in the previous sections. How can this novel scenario be construed in terms of IBE? Let us try to apply IBE on the case of string theory step by step. Already at the first step, the selection of theories, the situation does not quite match conventional cases of IBE: no ensemble of theories can be formed because only one candidate theory has been found. Indeed, contexts of this kind have been explicitly addressed by Alexander Bird (2007). He speaks of "inference to the only explanation" in those cases. In the absence of any other known explanations, one must infer that the only known explanation H is the viable theory. That kind of reasoning is schematized in Figure 3.1. A triangle signifies empirical input, a rectangle a theoretical statement (in the given case the theory H_c), a simple arrow an explanation and the broad arrow an inference to the best explanation (from a theory's explanatory character to its validity).

Since theory H is not empirically confirmed in the given case, we do not know whether H would still be a good explanation once all data is in, which means that IBE seems very weak. In order to strengthen it, we have to infer the probable viability of H on different grounds: we have to assess limitations to scientific underdetermination. That assessment once again takes the form of IBE, albeit at a meta-level. We understand that statements of limitations to scientific underdetermination (let us denote them Y) can explain the three kinds of observations discussed in the previous section: (a) the observation that scientists have not found any alternatives despite looking for them; (b) the

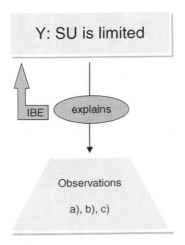

Figure 3.2 IBE at the meta-level.

observation that scientists developed the theory without considering a number of explanatory interconnections which emerged later on; and (c) the observation that in comparable cases within the research field scientific theories eventually tended to turn out predictively successful. Since we find no other equally satisfactory explanations of observations (a), (b) and (c), we infer the truth of Y. IBE at the meta-level of limitations to scientific underdetermination is represented in Figure 3.2. The trapezoid denotes the observations (a)–(c) which do not constitute empirical data in physics but are relevant for theory assessment. If sufficiently strong, the statement Y on limitations to scientific underdetermination now provides direct support for the viability of theory H and therefore assumes the role of a crucial element of IBE at the ground level. Non-empirical theory assessment thus involves IBE both at the ground level and at the meta-level, as drawn in Figure 3.3.

The presented analysis demonstrates that the concept of IBE can be extended to the case of non-empirical theory assessment. In fact, it even says a little more than that. It was mentioned in the beginning that assessments of underdetermination implicitly must enter IBE even in cases where empirical theory confirmation is available. Thus, the scheme I present may be seen as a completion for all IBE-type reasoning; to the extent that scientific reasoning is genuinely based on IBE, it must, at least implicitly, include a second level of IBE that allows for assessments of limitations to scientific underdetermination. In other words, assessments of scientific underdetermination may be more central to scientific theory assessment in general than one would assume at first sight. Part II of this book will further pursue that train of thought.

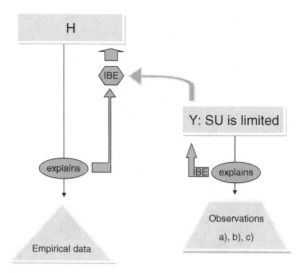

Figure 3.3 Overall inference scheme.

3.5 A Bayesian analysis

The currently most influential understanding of theory confirmation is based on Bayesian reasoning.[4] We have already encountered the Bayesian approach in Section 2.1. Theory confirmation in a Bayesian sense is provided by empirical data E that raises the probability of the truth (or, more generally, the viability) of theory H. As noted in Section 2.1, such empirical data is canonically understood to be constituted by data that is predicted (either deductively or statistically) by theory H. Therefore, it was argued, Bayesian theory confirmation can serve as an exemplification of the canonical paradigm of theory assessment. Does this imply that a Bayesian understanding of theory confirmation is at variance with non-empirical theory assessment of the kind presented in previous sections or is it possible to formalize non-empirical theory assessment along Bayesian lines of reasoning? Recent work by Dawid, Hartmann and Sprenger (in press) has analyzed this question and demonstrated that core procedures of non-empirical theory assessment do indeed constitute theory confirmation in a Bayesian sense. The following pages will not carry out a formal analysis but present the general layout of the argument.

Nothing in the Bayesian framework explicitly prescribes that data E must be predicted by theory H in order to constitute confirmation of H. The core reason

---

[4] Introductions to Bayesian epistemology in scientific reasoning are Bovens and Hartmann (2003) and Howson and Urbach (2006).

why one normally looks at data E that is predicted by H lies in the fact that the inequality $P(T|E) > P(T)$ that establishes confirmation (T once again being the statement that H is true or viable) can be proved straightforwardly in this case: if E follows deductively from H, then $P(E|T) = 1$. Since we cannot exclude other data than E in the absence of theory H, we have $P(E) < 1$, from which the above inequality follows based on Bayes' theorem. If E follows statistically, the situation is less rigid but similar. An extension of the Bayesian approach towards data that is not predicted by H must offer a strategy how to establish confirmation in the absence of the described simple line of reasoning. In fact, the discussion carried out in the previous sections has already provided us with the necessary tools to that end. It just remains to formalize them within the Bayesian framework.

Let us assume some evidence F of a kind that is not predicted by theory H. We call such evidence "non-empirical" with respect to H. Moreover, let us assume that F is predicted by another hypothesis Y. We thus can expect – under normal circumstances – that data F confirms hypothesis Y. Finally, let us assume that the truth of Y raises the probability of the truth of H. In that case, F raises the probability of the truth of H as well and therefore constitutes confirmation of H in a Bayesian sense.[5] In order to find a Bayesian formalization of non-empirical theory assessment, we therefore have to find a suitable candidate for hypothesis Y. The analysis of the previous sections suggests a specific candidate: a hypothesis on the number i of alternative theories that account for the available data and fulfil a given set of scientificality conditions. We introduce an infinite valued variable Y, where a value Y_i corresponds to the statement that there exist i alternative theories. Furthermore, we define Y_i^- as the statement that the number of possible alternative theories is lower than i. Y_i^+ denotes the inverse statement that there exist at least i alternatives. Simple statements which posit limitations to scientific underdetermination would be of the kind Y_i^-. More generally, a Bayesian formalization of a statement on limitations to scientific underdetermination attributes a probability $P(Y_i)$ to each Y_i.[6]

---

[5] This general strategy of theory confirmation has been discussed before in the context of the novel confirmation debate. See e.g. Maher (1988), Kahn, Landsberg and Stockman (1992), Barnes (2008) and Dawid (in press).

[6] There is a close relationship between assessments of limitations to scientific underdetermination and the "catch-all hypothesis" (Shimony, 1970), which is the hypothesis that we understand the spectrum of possible alternative theories sufficiently well to attribute a low value to $P(E|\neg T)$. (Low values of $P(E|\neg T)$ are required for having good confirmation by empirical evidence E.) A strong statement on limitations to scientific underdetermination can provide justification for low $P(E|\neg T)$ if one knows about the existence of any alternative scientific theory that does not predict E. Note, however, that the assumption of strong limits to scientific underdetermination is by no means the most common strategy of justifying low $P(E|\neg T)$. Low $P(E|\neg T)$ often results from simply assuming a very wide spectrum of possible alternative theories which do not predict E.

On that basis, a formalization of the argument of no alternatives (NAA) can be carried out. The relation between F and Y is most immediate. One can define a two-valued variable F_A with value F_A corresponding to the statement that scientists have not found any alternatives to theory H which are (expected to be) consistent with the available data and $\neg F_A$ corresponding to the statement that alternatives have been found. Now, one assumes, in accordance with the arguments of Section 3.1, that it is more likely that scientists do not find any alternatives to theory H the fewer the number of possible alternative theories that exist. The weakest way to formally implement this condition is to assume $P(F_A|Y_i^+) \leq P(F_A|Y_i^-)$ for all i and $P(F_A|Y_i^+) < P(F_A|Y_i^-)$ for at least one $i > 0$. It can then be shown that

$$\langle Y \rangle := \sum_{i=1}^{\infty} P(Y_i)Y_i > \langle Y \rangle_F := \sum_{i=1}^{\infty} P(Y_i|F_A)Y_i.$$

F_A thus lowers the expectation value of Y and thereby serves as an indicator for limitations to scientific underdetermination. It can also be shown that $\langle Y \rangle_F$ is finite even if $\langle Y \rangle = \infty$ for a large class of scenarios. Second, again following the line of reasoning in Section 3.1, one introduces rather weak formalized versions of the assumptions that both the viability of theory H and the occurrence of F are the more likely the lower the number of possible alternatives. Furthermore, one assumes that T is conditionally independent of F_A given Y. This amounts to the assumption that the belief in the empirical viability of H is not influenced by the information that scientists did not find alternatives if the number of possible alternatives is known exactly.[7] The Bayesian network representation of the given setup (including a variable D to be explained below) is given in Figure 3.4. Under the stated conditions, one finds

$$P(T|F_A) > P(T),$$

Figure 3.4 The Bayesian network representation of NAA.

---

[7] Strictly speaking, one would just need the very plausible assumption that F_A does not *increase* the probability that the unconceived alternatives are true. If F_A would *decrease* the probability of those theories being true, that would imply a further increase of $P(T|F_A)$ and thus just add another element of theory confirmation by F_A.

which means that F constitutes confirmation of theory H. In other words, under plausible assumptions NAA does amount to theory confirmation in a Bayesian sense.

The Bayesian analysis also shows the limits of NAA and elucidates the relatedness between NAA and MIA (the meta-inductive argument from predictive success in the research program). As already discussed in Section 3.1, the no alternative argument suffers from the fact that the complexity of possible solutions to the given scientific problem in conjunction with limitations to the scientists' capability might also explain the absence of known alternatives to theory H. If the complexity of the scientific problem is the true explanation of F, however, no inferences to the viability of H can be drawn. The problem is that F can never distinguish between an actual lack of alternatives and the scientists' insufficient capabilities. This threatens the significance of the no alternative arguments due to the entirely subjective character of initial probabilities. If we start with specific subjectively chosen prior probabilities for the specific statements Y_i^- and D_j^+ (where D_j parameterizes the complexity of the scientific question in comparison with the scientists' capabilities), future evidence F cannot change the ratio between these probabilities:

$$P(Y_i^-)/P(D_j^+) = P(Y_i^-|F_A)/P(D_j^+|F_A).$$

Thus, even though F_A formally constitutes theory confirmation, the probabilities of the viability of H extracted from F_A do not converge under future evidence F_A. The confirmation value of F_A therefore remains largely subjective. Once one believes that D_j^+ is the far more probable explanation of F_A than Y_i^-, no further observations F_A will ever constitute strong confirmation for Y_i^-. In order to remove this deadlock, we need evidence that supports Y_i^+ without supporting D_j^+.

MIA can provide such evidence. As argued in Section 3.1, predictive success within the research program can be understood as an indicator that scientific underdetermination within the research program tends to be limited. On the other hand, predictive success clearly does not suggest that scientists are not clever enough for finding viable theories in the research field. Therefore, calling data about predictive success in the research program F_M, we have

$$P(Y_i^-)/P(D_j^+) > P(Y_i^-|F_M)/P(D_j^+|F_M).$$

Increased probabilities $P(Y_i^-|F_M)$ extracted from the meta-inductive argument now can be used as priors for the argument of no alternatives. The meta-inductive argument thus is capable of strengthening the significance of the no alternatives argument.

The line of reasoning sketched above establishes two important points. First, ⟶s justified to call non-empirical theory assessment theory confirmation in a

Bayesian sense. And second, assessments of limitations to scientific under-determination provide a workable foundation for the Bayesian formalization of non-empirical theory confirmation. The core tenets of Chapter 3 thus are supported by a Bayesian analysis. Though the Bayesian approach at first glance seems to instantiate the canonical paradigm of theory assessment, in the end it turns out to imply the possibility and potential significance of non-empirical theory assessment.

PART II

A wider perspective

Chapters 2 and 3 have presented the characteristics of the mechanisms of non-empirical theory assessment which can be observed in the context of string physics. The question now arises whether the case of string physics is exceptional due to the specific context of its conceptual evolution and its particularly difficult empirical status. In order to address this question, let us start with a more general look at contemporary high energy physics.

4

The dynamics of high energy physics

4.1 Two crucial principles in modern high energy physics

Contemporary high energy physics[1] is based on two pivotal developments whose roots reach back more than half a century. Experimentally, it rests on the crucial role of deep inelastic scattering. Theoretically, it relies on internal symmetries as core structural characteristics of microphysics. Together, those two developments have generated the character of high energy physics as we know it today. Let me briefly discuss character and significance of those two developments.

Deep inelastic scattering

Early microphysics relied on various experimental methods for testing the structure and the constituents of atoms. Elastic scattering of particles on atoms, measurements of absorption and emission spectra and other experiments were deployed to test a wide range of atomic properties. From the 1940s onwards, a new strategy of experimental testing was added to the physicist's arsenal of experimental tools. Microphysics increasingly relied on deep inelastic scattering, the collisions of particles with very high kinetic energies which generated new particles in the process.

Conceptually, deep inelastic scattering is based on two fundamental principles of modern physics. On the one hand, special relativity establishes the equivalence between mass and energy. Each mass value corresponds to a certain energy value. Based on that equivalence, physical processes can involve the

---

[1] Classical textbooks on quantum field theory and theoretical particle physics are Weinberg (1996) and Peskin and Schroeder (1995). Books dealing with the history of high energy physics are Pickering (1984), Pais (1986), Riordan (1987), Galison (1987), Galison (1997) and Schweber 1994).

transformation of massive objects into radiation and vice versa in agreement with the principle of energy conservation. Quantum mechanics, on the other hand, introduces an irreducibly stochastic element to physics. It implies that particles can decay into other particles in agreement with energy conservation and the conservation of quantum numbers. The probabilities for such processes depend on coupling constants, which characterize interaction strengths between the involved particles, and on the spectrum of possibilities to realize the processes in phase space (i.e. the space of the involved particles' locations and momenta). Joining special relativity and quantum mechanics then implies that a collision of highly accelerated (and thus highly energetic) particles can turn the initial particles into all possible particle combinations whose production is consistent with the valid conservation laws. The experimental physicist therefore can determine the spectrum of particles that can in principle exist in our world by carrying out the appropriate scattering experiments.

There is an important constraint to the range of a collider experiment, however. Due to energy conservation, the particles produced in a scattering experiment can at most have a mass that corresponds to the overall collision energy. The collision energy attainable in an experimental apparatus thus sets a limit for the mass of particles which can be discovered. Since quantum mechanics attributes a charac- teristic wave length to each energy value (corresponding to the wave length of the free wave function of a particle with that energy), producing a particle collision with a certain collision energy corresponds to testing a certain distance scale as well. Testing higher energy scales thus corresponds to testing lower distance scales.

The described relation between collision energy and the capability to gen- erate new particles has set the path for experimental high energy physics for the last three-quarters of a century. Experimental particle physicists resorted to building particle colliders with increasingly high collision energies in order to look for particles with increasingly high rest mass. The length of the accelerator tube in the largest existing accelerator rose from 24 cm in 1931 to 5 m in 1942, 72 m in 1953 and 600 m in 1959. The Stanford Linear Accelerator of 1966 had a length of 3 km. The Large Electron Positron Collider (LEP) completed at CERN in 1989 measured 27 km in circumference. The LHC collider experiment, which is currently running in the 27 km tunnel built for LEP, took 16 years to build. About 8000 physicists work on the experiment and the analysis of its data. The experimental testing of fundamental micro-physics thus has turned from a small enterprise that could be carried out by an individual researcher into a gigantic effort of a large community of collaborating scientists and technicians.

Collider experiments roughly consist of two parts. First, there is the device that accelerates the particles (electrons and positrons, protons, antiprotons or heav ions) before they collide. This is done with strong electromagnets either in a circular storage ring or in a linear accelerator. Modern particle accelerators consist of several accelerating devices of increasing power which are used one after the other.

The second part is the detector surrounding the spot where the initial particles collide and the particles which have been produced in the collisions are identified. Deep inelastic scattering generates particles which cannot be observed in nature at the low energies which characterize our normal living environment because the conservation laws allow their decay into lighter particles. Only the lightest particles with a specific conserved quantum number are stable and constitute the building blocks of the world we observe. The larger the mass difference between an unstable particle and its decay products, the faster it decays, i.e. the shorter its average lifetime. Particle detectors can detect charged unstable particles with a sufficiently long lifetime by measuring the traces left by the particles' electromagnetic interaction with the detector. Particles with a very short lifetime do not move far enough to leave a visible trace on the pictures extracted from the detector. They only produce point-like vertices: the traces of those particles which have generated a short-lived particle and those which were generated after its decay meet at one point.

The interpretation of particle traces in terms of the involved particles is a highly complex affair. By generating an electromagnetic field in the detector, one can produce curved particle traces. The curve of a particle trace can then provide immediate information about the relationship between the corresponding particle's mass, velocity and charge. If the particle decays in the detector, the length of its trace can provide information about its lifetime. Uncharged particles can be inferred from gaps between traces or missing energies in the interpretation of the decay picture. The conjunction of all characteristics of the pattern of traces, vertices and gaps in the detector eventually leads to the attribution of traces and vertices to specific particles with some probability. A large number of collected events then can lead to a statistically significant statement regarding the existence of a specific particle type. The identification of particles on that basis constitutes the crucial experimental test of particle theories today. New theories predict a characteristic set of particles. If the existence of those particles can be extracted from the collected set of decay pictures at a certain level of statistical significance, this is considered a conclusive confirmation of the corresponding theory.

Gauge symmetries

Conceptually, today's high energy physics is based on the standard model of particle physics. The standard model was developed in the late 1960s and early

1970s in order to provide a coherent and full description of all nuclear interactions. Besides the electromagnetic interaction, nuclear physics had discovered two other kinds of interaction. The strong interaction was understood to be responsible for the binding of nucleons (and, when it was understood that the nucleons have constituents, for binding those constituents, the quarks). The weak interaction was necessary for understanding scattering processes which produced neutrinos, since the latter felt neither electromagnetic nor strong interaction.

In the 1960s, attempts to develop a theoretical understanding of weak and strong interactions were marred by a number of conceptual problems. A crucial problem was related to the question of the renormalizability of quantum field theory. (Quantum field theory is the relativistic generalization of quantum mechanics.) In the 1930s, it had been understood that the field theoretical calculation of particle scattering processes led to infinite terms. Feynman (1950) and Schwinger (1951) then developed the so-called "renormalization technique" which provided coherent calculations of electromagnetic interaction processes at the quantum level (thereby creating the research field of quantum electrodynamics (QED)) by canceling off the divergences in a physically meaningful way. It remained unclear, however, how to apply the renormalization technique to the other known microphysical forces, as there were strong and weak interactions.

The crucial concept deployed in the standard model for solving the renormalization problem is the concept of a gauge field theory.[2] Gauge field theory relates a certain kind of internal symmetry to the existence of a specific set of spin 1 particles, the so-called gauge bosons. The simplest type of a gauge theory, called abelian gauge theory, only posits one single gauge boson. More complex, so-called non-abelian gauge theories posit larger gauge symmetry groups and, correspondingly, a higher number of gauge bosons. Such non-abelian gauge theories are based on the existence of a specific form of symmetry under transformations in a space of particle degrees of freedom. In other words, if one replaces one kind of particle by another that is related to the first by the symmetry, the physics of the system does not change.

Already in the 1950s, it was suspected that the renormalizability of QED could be explained by the fact that it had the form of an abelian gauge theory: the photon played the role of a gauge boson. Yang and Mills (1954) suggested that the strong interaction might be based on the exchange of a set of vector bosons. The resulting theory then should be characterized by a generalization of QED's abelian gauge symmetry, namely a local and continuous permutation symmetry

---

[2] Philosophical perspectives on gauge field theory are presented e.g. in Brading and Castellani (2003).

between the types of elementary particles bound together by the interaction.[3] In analogy with the electromagnetic case, this non-abelian gauge symmetry could then enforce renormalizability just like the abelian gauge symmetry had done in the case of QED. A few years later, the idea of Yang and Mills was used for developing a unified picture of electromagnetic and weak (i.e. electroweak) interactions by Glashow (1961) and Salam and Ward (1964) and for describing the strong interaction based on the quark picture (see below) (Han and Nambu, 1965).

The gauge theoretical approach faced one serious problem, however. Gauge symmetry implied that vector bosons had to be massless and all fermionic particles connected by a gauge symmetry had to have the same mass. Both conditions were incompatible with the observed empirical data on weak interaction. The mechanism of spontaneous symmetry breaking (Goldstone, 1961; Nambu and Jona-Lasinio, 1961; Higgs, 1964) provided a solution to this problem. It conjectured a new scalar particle, the Higgs particle, whose interaction structure had the effect that the ground state of the theory broke the gauge symmetry even though the Lagrangian of the theory was gauge symmetric. This "spontaneous symmetry breaking" allowed for a realistic mass spectrum of fermions and gauge bosons. The resulting mass of the Higgs particle itself determined the energy scale at which spontaneous electroweak symmetry breaking took place, the so-called electroweak scale that has already been referred to at several points in this book. In 1967, Weinberg finally integrated all these developments in a coherent way in what is today called the standard model of particle physics and thus provided a fully adequate conception of the electroweak interaction (Weinberg, 1967). 't Hooft's proof of the renormalizability of non-abelian gauge theory ('t Hooft, 1971; 't Hooft and Veltman, 1972) finally established the standard model as the most promising candidate for a viable description of electroweak interactions.

A second fundamental problem faced by particle physics in the 1960s was related to some surprising features of nucleons and other heavy fermionic particles (called Hadrons). An increasing number of such Hadrons of different types were found in collider experiments during the 1950s and 1960s. The vast number of such particles, which were at the time all taken to be elementary, was highly confusing and drew into question the significance of looking for fundamental particles in general. In 1964, Gell-Mann realized that the charges and masses of all those Hadrons were compatible with interpreting them in terms of boundstates of two or three constituents, which he called quarks (Gell-Mann, 1964). In 1969, Feynman was the first to understand that a curious and so far

---

[3] Yang and Mills at that time still took the nucleons to be elementary particles.

unexplained feature of scattering processes with nucleons at medium energies (the so-called Bjorken scaling) could be understood as an indication that nucleons had constituents which behaved like free particles at those energy scales (Feynman, 1969). Joining the two ideas led to the coherent conjecture that Hadrons consisted of constituent elementary particles. Something about those quarks was deeply confusing, however. If they were actual particles, why had no one observed them in isolation? And if the force binding them together was so strong that they could not be separated, why did they behave like free particles as long as they remained close to each other (which had to be assumed to explain Bjorken scaling)? In the early 1970s it was understood that gauge field theory could provide an answer to those questions. Politzer (1973) and Gross and Wilczek (1973) showed that non-abelian gauge theory possessed some interesting and unusual properties. Its couplings became very weak at small distances (a property called asymptotic freedom) and exponentially strong once the interacting particles moved away from each other (a property called confinement). These properties which could be deduced on a theoretical level provided an excellent explanation for quark phenomenology. They explained the short distance free movement of quarks as well as the impossibility to isolate them. Gauge theory thus became a convincing candidate for the description of strong interaction as well.

In 1973/74, the standard model of electroweak and strong interactions appeared convincing because it was the only available candidate for solving a wide range of puzzles in high energy physics in a coherent way:

- It offered a solution to the crucial technical problem of renormalizability of strong and weak interactions.
- It explained the phenomenon of asymptotic freedom of the quarks under strong interaction.
- It allowed for different masses of matter particle types.
- It provided a viable system of quark bound states that could explain the large number of fermionic particles found in the 1950s and 1960s.

The introduction of gauge theories is most significant, however, due to the new level of empirical predictions it offered. The character of those predictions differs significantly from the character of earlier predictions in microphysics. Before the advent of gauge theory, microphysical theories were capable of determining the dynamics of microphysical objects. Which kinds of elementary objects existed, however, remained to be determined by experiment.[4] As

---

[4] There were some early requirements on the types of existing particles. In particular, relativistic quantum theory required that for each particle there existed an anti-particle.

mentioned above, a wide range of new particles was discovered in collider experiments during the 1950s and 1960s. At the time, none of them could be derived from basic physical principles. The introduction of new fundamental particles, the quarks, constituted one important step towards explaining the spectrum of observed particles. By constructing all possible quark bound states, one could explain the observed fermionic particles and could even predict new ones which had not yet been discovered. Gauge theory went one substantial step further. Even the spectrum of fundamental particles could now be derived to a large extent from more fundamental structural characteristics of the theory, from the gauge symmetry structure. The existence of a gauge symmetry between particles required that the existing particle types formed complete multiplets of representations of the gauge symmetry groups. The particle spectrum therefore was strongly constrained by the gauge symmetry. Particle ontology became theoretically predictable. Since a gauge theoretical structure is required for conceptual reasons in order to make the overall conceptual framework of particle physics coherent, the prediction of particle types is a good example for the increasing power of consistency arguments in microphysics.

What were the specific predictions of the standard model? First, it predicted further Hadronic particles which were now understood in terms of quark boundstates. Beyond that, and more significantly, it also predicted a number of new elementary particles and relations between the couplings of particles based on the rigid requirements of the theory's internal symmetry structure. In this vein, the standard model predicted the existence of W- and Z-Bosons which constituted gauge Bosons of the electroweak interactions; the existence of gluons which constituted the gauge Bosons of the strong interaction; the existence of new quark and lepton types for symmetry reasons and in order to account for the observed CP-violation (Kobayashi and Maskawa, 1973); and finally the existence of a Higgs Boson responsible for spontaneous symmetry breaking. All those particles were understood to be testable in collider experiments of sufficient size but none of them had been discovered before 1974.

In the following two decades, new large colliders built in order to detect the particles predicted by the standard model led to the experimental discovery of most of the elementary particles predicted by the standard model. The charm quark and tauon were found in 1974; the bottom quark in 1977; gluons in 1979; the W-Boson in 1983; the Z-Boson in 1984; and the top quark in 1994. The Higgs Boson was finally discovered during the LHC experiments at CERN in summer 2012.

If we look at the state of particle physics in the 1970s and 1980s, we thus observe strong dynamics at both the experimental and the theoretical level, which jointly propelled the development of the field. Experimentally, the collider technology offered a continuously applicable strategy for testing increasingly high energy scales. Conceptually, the deployment of symmetry principles – and of gauge symmetries in particular – provided a basis for theoretical development. Symmetries played an intricate double role in this process. On the one hand, they turned out to be instrumental for solving crucial consistency problems in high energy physics. On the other hand, they constrained the freedom of model building in several ways and thus generated strong predictions. The latter effect has been strong enough to lead to an inversion of the role play between theory and experiment. While up to the 1960s experiment was to a large extent deployed for discovering new phenomena which then had to be interpreted theoretically, the 1970s and 1980s were mostly characterized by experimental tests of theoretical predictions. Theory had taken the lead in the field's dynamics, but theory and experiment were still progressing side by side.

What happens from that period onwards can be related to the different fates of the two described driving principles behind progress in particle physics. Collider physics was less and less capable of keeping pace with the theoretical predictions. At the same time, the use of high level mathematical tools like gauge symmetries or other newly emerging theoretical principles continued to flourish and to generate new conceptual progress. The discrepancy between a slowed down dynamics of the guiding principle of experimental high energy physics and an unbroken dynamics of theory development has created the empirically problematic situation faced by fundamental physics today. In this light, the precarious empirical status of contemporary theories appears to be related to the general mechanism that has carried progress in high energy physics over the last half of a century. The problem reaches far beyond string theory and must be faced by all contemporary theories in fundamental physics.

4.2 Theory dynamics beyond the current limits of empirical testability

In order to bolster the previous statement, let me briefly discuss the most influential theories in fundamental physics today.

It is important to remember that scientific theories whose characteristic predictions have not yet been empirically confirmed nevertheless rest on empirical data. Either they account for known empirical data that had remained unaccounted for up to that point or they solve a theoretical problem that has arisen in previous theories about the known empirical phenomena. The difference between an empirically confirmed and an empirically unconfirmed theoretical claim lies in the path of reasoning that leads from the empirical data to holding the claim in question. According to the simplified canonical under-standing of scientific theory assessment, an empirically confirmed claim can be accepted based solely on the coherent interpretation of the known empirical data. To the contrary, trusting an empirically unconfirmed scientific claim relies on a more complex line of reasoning. The theory implying the claim is consid-ered a candidate for a viable theory about nature because it provides a theoret-ical solution to a scientific problem that arises based on the available empirical data. Trust in the claim rests on the belief that no scientifically satisfactory solution to the given problem exists that does not imply the theory's validity. In other words, trust is based on an assessment of limitations to scientific underdetermination.

As emphasized above, the lack of empirical evidence beyond the standard model has not resulted in a slowdown of theory building and a lack of clear perspectives regarding novel theoretical approaches in high energy physics. The strategy of developing new theories based on principles of universality and consistency has been fertile even in the absence of empirical tests. From the mid 1970s onwards, a series of new concepts has entirely reshaped our understand-ing of theoretical microphysics. Those new concepts were developed for a number of theoretical reasons as well as in order to provide a more convincing explanation of the observed data. None of them, however, was based on incompatibilities of empirical data with standard model predictions. In the following, the most important of those theories beyond the standard model shall be briefly presented.

Grand unified theories

Grand unified theories (GUTs), the first of which was presented in Georgi and Glashow (1974), have played a leading role in particle physics model building for many years. The key idea of grand unification is a rather technical one. The gauge structure of the standard model consists of three simple Lie groups (SU(3), SU(2) and U(1)), which, roughly speaking, correspond to the three gauge interactions: strong, weak and electromagnetic. Each of the three gauge interactions is characterized by an individual gauge coupling which determines the interaction strength. The values of the gauge couplings are quite different from each other at the energies which can be tested in collider experiments today. However, due to quantum corrections the values of the gauge couplings are dependent on the energy scale at which they are measured. In other words, particles colliding with a very high collision energy experience substantially different coupling strengths than particles colliding at lower energies. (We have encountered this so-called running of the gauge coupling already before. It is responsible for confinement and asymptotic freedom of particles bound by the strong interaction.)

Now it turns out that all three gauge couplings assume roughly the same value at some very high energy scale, the so-called GUT scale (at energies 10^{12} times as high as those testable by present collider experiments). The fact that all three couplings meet roughly at one scale is non-trivial and may be taken as an indication for the existence of a grand unified structure of all gauge interactions. At the GUT scale and beyond, all gauge interactions then could be described by one simple gauge symmetry, i.e. a symmetry characterized by one simple Lie group.[5] Below the GUT scale, however, that symmetry must be spontaneously broken, which means that it is realized in the Lagrangian itself but is not realized in the ground state of the theory. Below the GUT scale, only the three smaller symmetries $SU(3) \times SU(2) \times U(1)$, which correspond to the three nuclear interactions, remain intact. (The electroweak symmetry $SU(2) \times U(1)$ is eventually broken down to the electromagnetic $U(1)$ at the electroweak scale, which is testable by present-day collider experiments.)

Two core motivations thus lead to the hypothesis of grand unified theories. The first one is empirical. The measured values of the three gauge couplings imply that all of them roughly meet at one scale. While such a constellation looks like a mere coincidence in the context of the standard model, grand unified theories can explain it. The second motivation is of an entirely theoretical kind. One universal coupling constant at a fundamental level looks more satisfactory than three of them.

The conjecture of a grand unified theory implies the existence of a large number of additional gauge vector particles and a large number of new matter particles. Those additional particles result from the large representations of the larger grand unified gauge symmetry group that has to be constructed in order to account for the interaction structure we observe at low energies. All those new particles, however, have a mass of the order of the GUT scale and are therefore far too heavy to be observed by collider experiments. Direct empirical confirmation of the theory's core predictions thus cannot be expected within the framework of collider experiments.

Indirect empirical evidence for or against grand unified theories can be provided by the search for proton decay. Proton decay would univocally constitute a signature of new physics beyond the standard model, since the latter predicts that the proton is absolutely stable. If proton decay was discovered, this would constitute considerable support for the GUT concept. It

---

[5] It turns out that only three such grand unified groups are compatible with the data: SU(5), SO(10) and E(6).

would be difficult, however, to link proton decay univocally to a GUT without observing GUT-particles themselves since it could be induced by other theories beyond the standard model as well. Vice versa, a failure to observe proton decay in measurements with increasing accuracy reduces the parameter space for possible GUT scenarios. The fact that proton decay has not been discovered up to now already rules out the simplest GUT theory (minimal SU(5)) and significantly reduces the allowed parameter space for supersymmetric minimal SU(5) GUTs. Much leeway remains, however, for avoiding proton decay of a measurable size in more complex GUT scenarios, which makes the GUT concept practically irrefutable by proton decay measurements in the foreseeable future.

Supersymmetry

A theoretical concept that actually may be empirically confirmed at the LHC is supersymmetry. The basic idea is, once again, related to the theory's symmetry structure. So far, we have encountered internal symmetries, i.e. symmetries under rotations in the space of particle types, as the crucial symmetries in gauge field theory. In the early 1970s the question arose as to whether the Lie group structure defining internal symmetries is the most general form of a continuous symmetry that involves internal degrees of freedom. It turned out that the maximal consistent generalization of such an internal symmetry is supersymmetry, a symmetry between particles of different spin based on an intertwining of the inner symmetries so crucial in gauge field theories and the Lorentz symmetries which characterize special relativity. A supersymmetric quantum field theory in four dimensions was first formulated in Wess and Zumino (1974). If the world were supersymmetric, this would imply that each of the particles known today would have a so-called superpartner, i.e. a corresponding particle with different spin. The fact that no superpartners have been observed so far once again would have to be explained by a spontaneous breaking of supersymmetry at some higher energy scale.

In principle, the scale where supersymmetry is broken might be very high up, close to the GUT scale or beyond. However, two arguments suggest a supersymmetry scale that lies quite close to the electroweak scale. The first argument is related to GUTs. As we have mentioned above, GUTs are based on the observation that the three running gauge couplings meet at one point. This, however, is only roughly true in a standard model context. It turns out that the couplings meet much better in a supersymmetric scenario where supersymmetry is broken at a rather low energy scale. (However, supersymmetry is not the only possible extension of the standard model that can provide this result). The second argument is related to the strangely large scale difference between GUT and Planck scale (which is the scale where the gravitational interaction becomes as strong as the nuclear interactions) on the one hand, and the scale of electroweak symmetry breaking on the other. For technical reasons, supersymmetry can provide an explanation of that scale difference based on small integer factors which can have exponential implications. The scale difference of 10^{12} thus could be explained by a factor 12 somewhere in the formula, which may be produced naturally. Since the mechanism only works as long as supersymmetry is unbroken, the idea that supersymmetry creates the large scale difference would imply that no large scale differences between the supersymmetric scale and the electroweak scale exists.

Both arguments thus suggest that supersymmetry might be found at the LHC. If so, that would constitute the first substantial empirical step beyond the standard model.

Theory assessment of low energy supersymmetry and grand unified theories

Unlike string theory, supersymmetry and GUTs are supported to a certain degree by quantitative characteristics of the observed data. As discussed above, both theories gain credibility from the measured values of the coupling constants of strong, weak and electromagnetic interactions. However, the measurable core characteristics of both theories (in both cases constituted by a characteristic spectrum of new particle types) lie beyond the reach of empirical testing today. Therefore, the question of the significance of the empirical corroboration is still closely related to assessments of underdetermination based on the no alternative argument (NAA): can physicists conceive of alternative conceptual scenarios that reproduce the empirical signatures which are taken to support a specific theory? Such considerations indeed play an important role in the physical analysis and are capable of changing the overall assessment with time. In the case of GUTs, two significant instances of a broadening of the perspective have occurred. First, string theory has opened up a broader understanding of the implementation and breaking of symmetries based on the compactification of extra dimensions. Second, it was later understood that large extra dimensions could imply scenarios of gauge coupling unifications which differ significantly from the canonical understanding of grand unification. Both of those developments led to the acknowledgement that the spectrum of possible explanations of the observed pattern of gauge couplings was wider than initially expected. Exponents of GUTs in the late 1970s may have assumed that the empirical data on the values of gauge couplings allowed for a certain degree of confidence that the basic mechanism of grand unification as it was understood at the time was viable. Today's broader understanding of the conceptual context no longer supports this understanding but enforces a more general notion of what the data suggests.

In the case of supersymmetry, assessments of scientific underdetermination are also considered. As described above, low energy supersymmetry makes the meeting of the three running gauge couplings at one point more precise. Based on the assumption that a unification of gauge couplings is indeed the case, this feature provides one reason for believing in the viability of low energy supersymmetry. However, low energy supersymmetry is not the only way to improve the meeting of the gauge couplings. The introduction of other types of new particles could achieve similar results. The viability of low energy supersymmetry thus cannot be argued for based on NAA alone. In order to make a reasonable case, one has to resort to other kinds of reasoning. Low energy supersymmetry not only improves gauge coupling unification, it also offers a basis for explaining the large scale difference between electroweak scale and the GUT scale. Moreover, supersymmetry offers a coherent integration of gravity into a gauge field theoretical framework and is predicted by string theory. All those features had not been known at the time supersymmetry was developed in the early 1970s. Supersymmetry thus provides a nice example of a theory that offers unexpected explanations and interconnections in the sense of UEA. It is only on that basis that the unification of gauge couplings can become a meaningful argument; while supersymmetry is just one among many concepts that does the job of improving unification, among all the concepts scientists could think of low energy supersymmetry seems most convincing in the light of UEA.

Like in the case of string theory, the empirical vindication of the particle physics standard model (MIA) provides the background for a certain degree of trust in NAA and UEA in the given context. The argument does not show the same strength as in the case of string theory, however. In the latter case, the parallels with the standard model are stronger in an important respect: string theory, like the standard model, constitutes a conception that provides a solution to outright consistency problems in the given context. The standard model offered a renormalizable theory of nuclear interactions; string theory seems to offer a coherent unification of nuclear interactions and gravity. Arguments of no choice in the given case mean that no coherent alternative theory in a certain regime is available. MIA then can be deployed for the meta-inductive argument that, given that this kind of no choice argument was empirically vindicated in the case of the standard model, it may be expected to be valid in the case of -ing physics as well. Low energy supersymmetry and GUT are not required in order to make the theoretical description coherent. They merely serve to explain certain conspicuous quantitative aspects of the measured data: the values of the gauge couplings and the hierarchy between electro-weak scale and GUT/Planck scale. Thus, an additional question arises: is an explanation of the given relations necessary and what kind of explanation would one accept? This additional uncertainty makes the inference to the viability of the theories in question less stable.

Compared to the case of string physics, scientists working on low energy supersymmetry or GUTs thus can rely on more specific empirical predictions and more specific empirical corroboration, but, if anything, are less sure about their theory's viability. Neither the arguments for low energy supersymmetry nor the theoretical arguments for unification or the arguments for schemas of grand unification have been taken to be as cogent and forceful as the overall case that can be made for the viability of string physics. Therefore, no fundamental debate on the scientific legitimacy of the involved pattern of reasoning has been provoked by the way scientists working on GUTs and supersymmetry were resorting to the strategies of non-empirical theory assessment. Though some more conservative phenomenological particle physicists do not feel at ease with what they take to be the unabashedly speculative character of the described theories, the theories' proponents are mostly understood to adhere to the canonical distinction between establishing scientific knowledge about the world based on empirical confirmation and making "subjective" assessments of a theory's chances of being viable based on theoretical considerations: they deploy strategies of theory assessment in order to decide whether continued work on the theory seems promising with respect to possible future empirical confirmation.

It is important to note, however, that the strategies deployed in the described cases are of the same kind as those deployed in the context of string theory. Just like in the case of string theory, arguments of non-empirical theory assessment are piled up in several layers. Theoretical multi-step arguments like the one that leads from the observed pattern of gauge couplings to the existence of grand unification and from there to increased trust in the viability of supersymmetry can only be taken seriously if one has considerable trust in the cogency of the involved kind of reasoning at each individual step of the argument.

Moreover, it is important to consider the time scales involved. Theoretical analysis based on light empirical corroboration has constituted the only basis for theory assessment regarding the discussed theories for several decades. Therefore, it has naturally assumed a more important role than in other fields of physical research. Looking at the situation from the perspective of string theory, the role played by non-empirical theory assessment in theories like supersymmetry and GUT constitutes a test case for the next step in the empowerment of non-empirical theory assessment in string physics.

Supergravity

Soon after the construction of the first supersymmetric models, it became clear that a formulation of supersymmetry as a local gauge symmetry (= supergravity) had the potential to provide a fuller understanding of the particle character of gravity (Freedman, Ferrara and van Nieuvenhuizen, 1976). The particle that corresponds to the gravitational force in a field-theoretical formulation of gravity, the so-called graviton, naturally arises as the superpartner of the gauge particle of supersymmetry. Perturbation theory of supergravity then can be understood in terms of graviton-induced corrections to a flat background spacetime. For a while, supergravity was considered a good candidate for a fundamental unifying theory of gravity and the nuclear interactions. In particular, it was hoped that extended forms of supergravity could be finite and thereby solve the problem of the non-renormalizability of quantum gravity. Failures to establish the finiteness of supergravity and some arguments against that possibility have later led to a perspective where supergravity plays the role of an effective theory of string theory. Supergravity today is so closely related to string theory at a conceptual level that assessing the theory's status has by and large become a part of assessing the status of string theory.[6]

Cosmic inflation

Finally, one should mention a theory that has been developed in the context of cosmology and has in recent years contributed to a growing rapprochement between the fields of high energy physics and cosmology: cosmic inflation. Inflationary cosmology was first suggested in the early 1980s by Alan Guth (1981) and developed further by Andrei Linde (1982) as a way of explaining two conspicuous properties of the universe as we observe it: the isotropy and homogeneity of the cosmic microwave background, which denotes the fact that the cosmic background radiation measured by us is with considerable precision the same from whatever direction it comes; and the approximate flatness of spacetime. Both properties seem inexplicable within the framework of a traditional conception of the dynamics of the universe based on the principles of general relativity and quantum physics. That framework would imply that different parts of the universe observable by us today were causally

---

[6] As already mentioned in Section 1.3, recent work by Bern, Dixon and Roiban (2007) has suggested that $N = 8$ supergravity might be a finite theory after all. If so, it might play a role independently from string theory and, in fact, could in principle even be seen as an alternative to string theory. While the issue is undecided at this point, it can in any case be taken as an example for the volatility of arguments of limitations to scientific underdetermination.

unconnected at the earliest stages of the universe. If that were the case, however, physics as we know it could not offer any mechanism that would enforce the fact that all those previously causally unconnected parts of the universe have evolved in a way that makes them look so similar to each other in our experiments. The observed flatness of spacetime, on the other hand, would require an enormous fine-tuning of cosmological parameters for which conventional cosmology offers no explanation. Inflation solves both problems by introducing a mechanism that enforces a phase of exponential growth at a very early stage of the universe. That phase pulls causally connected parts of the universe so far apart that our present observational horizon only includes parts of the universe which were causally connected before the inflationary phase. Furthermore, the inflationary phase "flattens out" spacetime so that it explains the flat spacetime we observe without having to rely on fine-tuning.[7]

The established techniques of measuring the universe clearly imply that the universe we observe exited the inflationary phase many billion years ago to enter a phase of slower expansion. Therefore, a consistent theory of inflation requires a mechanism that explains the transition from the inflationary phase to the phase of normal expansion we observe. The analysis of inflation has led to two conclusions in this respect. First, no mechanism seems available that could trigger the phase transition of an entire exponentially expanding universe at the same time. Second, such a complete transition is not indicated by the empirical data anyway. The empirical data are fully consistent with a scenario where the part of the universe observable by us started its normal expansion phase at some point as a minimal patch of the size of the Planck length while the rest of the universe continued its inflationary expansion.

On that basis, the multiverse scenario of eternal inflation was developed, which today constitutes the standard approach to inflationary cosmology. According to eternal inflation, there exists a background of eternally inflating (i.e. exponentially expanding) space. Within this background space, quantum oscillations can, with a certain probability, trigger a phase transition to a phase of slower expansion at some point. At that point, a universe of slower expansion starts and expands within the inflationary background according to its own dynamics. The starting point of such a universe is perceived by observers within

---

[7] Inflationary cosmology in fact was first developed by Guth as a means of explaining another phenomenon, the observed lack of magnetic monopoles in a grand unified framework. Within a conventional cosmological framework, grand unified theories predict that we should see a large number of magnetic monopoles as remnants from very early phases of the universe. No magnetic monopoles have been observed, however, which was considered a serious threat to the idea of grand unification. Inflation can dilute the density of magnetic monopoles to a sufficient degree to explain their absence in empirical data. Inflation thus constitutes a nice example of unexpected explanatory interconnections, which arguably enhance the trust in that theory.

it as a big bang. Processes beyond the limits of that universe remain unobservable for such observers. Since the creation of a new universe with non-inflationary expansion is a matter of quantum statistics, they occur time and again at different points in inflationary background space. The overall scenario thus implies a multiverse, a huge, maybe infinite number of universes in an inflationary background space.

Cosmic inflation has been developed based on empirical data on isotropy and flatness of the universe. In recent years it has found some further empirical corroboration. In particular, recent precision measurements of large scale anisotropies of cosmic background radiation have been found to be in good agreement with models of cosmic inflation. Still, the collected data cannot be considered to be conclusive empirical evidence for cosmic inflation. On the one hand, cosmic inflation would not have been refuted even by substantially different data, since the fine-tuning of constants could produce models with substantially different phenomenology. On the other hand, it is not sufficiently clear to what extent so far unconceived alternative explanations could reproduce the collected data as well. Eternal inflation is a particularly interesting case because it predicts other universes which are inaccessible to us as a matter of principle. Core predictions of the theory thus lie fundamentally beyond the range of experimental confirmation. Reasoning for eternal inflation must be based on the general argumentation for inflation joined with consistency arguments indicating that eternal inflation constitutes the only coherent way to construct a theory of inflation. In other words, assessments of the status of inflationary cosmology will most probably have to rely heavily on assessments of scientific underdetermination for a long period of time.

Loop quantum gravity

Up to now, all presented theories were part of or largely influenced by the high energy physics research program. All those theories showed strong mutual interconnections and may be understood to constitute one overall perspective on fundamental physics. The theory to be presented now has been developed on a different basis and is understood by some of its exponents as an alternative to the former research program. Loop quantum gravity stands in the tradition of canonical quantum gravity, which aims at developing a quantized version of general relativity. The basic idea of loop quantum gravity is to generate space-time by building it from "quanta of space" which represent its minimal constituents. Technically, the approach is based on spin-networks, which constitute the web of nodes which represent the quanta of spacetime. The research program of loop quantum gravity was started in 1986 (Ashtekar, 1986) and has attracted considerable interest since then (Rovelli and Smolin, 1990; Rovelli, 1998). Today, it constitutes by far the most popular approach in canonical quantum gravity.

Loop quantum gravity faces the same difficulties to achieve empirical testing as the previously mentioned theories. In fact, its status can be compared to the cases of string theory and supergravity which haven't even found indirect empirical corroboration. Twenty-five years after it was first proposed, loop quantum gravity remains empirically entirely unconfirmed and has not developed a clear perspective for empirical testing in the foreseeable future. Conceptually, a number of crucial technical problems are still unsolved at this point. Most significantly, it is not quite clear whether loop quantum gravity has general relativity as its classical limit (though this is suggested by circumstantial evidence). Loop quantum theory thus remains a conceptually incomplete theory like string theory or cosmic inflation.

The assessment of the chances that loop quantum gravity constitutes an empirically viable theory thus must rely on non-empirical theory assessment just like the other cases mentioned in this section. Leading exponents of loop quantum gravity tend to be fairly confident regarding their theory's chances of success. However, compared to the way the status of string theory is assessed by that theory's leading exponents, it may be fair to say that assessments of loop quantum gravity are significantly more timid. While the sentiment that their theory is "too good to be false" is widespread among string physicists, adherents to loop quantum gravity are more inclined to emphasize that their approach constitutes one possible solution that has some attractive features but may well turn out false in the end. The question thus arises whether this difference in attitude can be explained based on the different degrees of applicability of the three introduced strategies of non-empirical theory assessment. A detailed in-depth study of this question would go beyond the scope of the present book. A brief sketch of some arguments shall be given, however, not so much with the aim of finding a conclusive verdict but rather in order to exemplify the way comparisons of the significance of non-empirical theory assessments in various contexts may be carried out.

If one tentatively tries to evaluate the degree to which the three non-empirical arguments apply to loop quantum gravity, one indeed finds that, while they can be applied to a certain extent, they seem weaker than in the case of string theory.

The difference is most conspicuous in the case of NAA, the argument of no alternatives. As discussed in Chapter 2, string theory may, with a slight caveat due to some new ideas which are currently developing, be called the only promising approach for a full unification of all interactions. Loop quantum gravity is no candidate for a full unification of the theories of particle physics and cosmology in its present state but just aims at reconciling gravity with the principles of quantum mechanics. In the latter enterprise, however, string theory obviously is a serious contender. If string theory was a viable theory of all interactions, it would necessarily provide a viable reconciliation between general relativity and quantum physics as well.[8] One thus observes an asymmetry between string theory and loop quantum gravity with respect to NAA. While loop quantum gravity at this point does not seriously infringe on the no alternatives claim regarding string theory, string theory does so with respect to loop quantum gravity. In addition, there are a number of alternative approaches in canonical quantum gravity which, though facing serious problems, may not be ruled out entirely as contenders of loop quantum gravity. To conclude, while loop quantum gravity clearly operates in a field where successful solutions are very difficult to come by, there are alternative approaches which must be taken seriously and therefore prevent a powerful application of the no alternatives argument.

Let us next look at UEA, the argument of unexpected explanatory interrelations. Loop quantum gravity has indeed produced interesting results which had not been foreseen when the theory was first developed. For example, it has turned out that loop quantum gravity provides an intuitively appealing picture of the final contracting phase of a re-contracting universe. In a simplified scenario, it dissolves the final singularity of classical general relativity in a way that implies a bounce-back of the universe from its final collapse. Loop quantum gravity also gives a qualitative microphysical explanation of black hole entropy. It may be argued, however, that those results are less convincing as exemplifications of unexpected explanatory interconnections than those provided by string theory. We remember that, in the case of string theory, very general structures like supersymmetry or supergravity, which had been appealing on their own grounds, emerge naturally as a necessary implication of string theory. The dissolution of the initial/final singularity, and the resulting bounce-back universe, to the contrary, may seem appealing (and some form of dissolution of spacetime singularity clearly is required in a theory of quantum gravity) but do not really connect to a well-motivated but seemingly independent context of theory building. Loop quantum gravity, to make the argument

---

[8] Proponents of loop quantum gravitation make the claim that the full background independence built into their approach (that is, the property that no spacetime structure has to be introduced first in order to start calculations) gives it a substantial advantage over string theory. Without entering the details of that argument, it suffices for the present purpose to stress that (i) its cogency is not agreed upon by all participants to the debate and (ii) supporters of the argument do not insist that the argument conclusively rules out string theory. Therefore, string theory must count as an alternative to loop quantum gravitation even for the proponent of the latter.

more general, does not play the role of a theory that interconnects a number of independently well-motivated fundamental theories in the way string theory does. Regarding the explanation of black hole entropy, loop quantum gravity can explain the proportionality between entropy and event horizon and can be applied coherently for various kinds of black holes. The argument thus is not restricted to near-extremal black holes like in the case of string theory. Unlike string theory, however, it does not explain the correct factor of black hole entropy, which has to be inserted by hand. On these grounds, the explanation of black hole entropy by loop quantum gravity arguably looks less striking than in case of string physics.

MIA, the meta-inductive inference from empirical success within the research field or research program, also seems to give less support to loop quantum gravity than to string theory. Both loop quantum gravity and string theory are situated within the general context of general relativity and quantum physics and therefore can use the predictive successes in these fields in the sense of MIA. The context, however, where predictive success has been most striking in last 50 years has been the standard model of particle physics. String theory, which is largely based on the physical concepts deployed in the standard model, profits decidedly more strongly from these successes via MIA than loop quantum gravity, which constitutes a research program that is fairly independent from that field.

The analysis given in the previous paragraphs exclusively addresses those forms of non-empirical theory assessment that are based on arguments of limitation to scientific underdetermination. Therefore, it cannot offer an overall assessment of the chances of success faced by loop quantum gravity, which may also be supported by other lines of reasoning. The analysis demonstrates, however, that it is justified to differentiate between the status of loop quantum gravity and string theory based on the question limitations to scientific underdetermination.

4.3 Theory assessment in a new scientific environment

The examples provided in the last section show a clear pattern. Fundamental physics has entered a stage in recent decades where the empirical evidence for new theoretical conceptions is scarce or remains entirely absent for many decades. It is often not clear whether conclusive empirical testing of a theory can be achieved at all or whether piecemeal empirical corroboration and circumstantial evidence is the maximum one can hope for. Those circumstances alter substantially the parameters of theory assessment and raise important new questions for the philosophy of science. As long as the phases of inconclusive or absent empirical evidence were short in most cases, a detailed analysis of a theory's status during its unconfirmed state seemed unnecessary. Theory assessment could reasonably focus on empirically tested theories. The most interesting aspect of theory assessment was empirical confirmation. Questions to be addressed in that vein by philosophers of science were the mechanisms and the rationale of theory confirmation, the connection between empirical confirmation and truth, the nature of theory succession from one empirically confirmed theory to another and other related questions. Theories that were not confirmed empirically remained largely outside the scope of philosophical analysis. Such theories were considered speculative hypotheses whose generation constituted a crucial element of the research process, whose assessment, however, could wait until they had been either refuted or confirmed by empirical data.

In the new context of particle physics, the exclusive focus of the philosophy of science on empirically confirmed theories does not seem satisfactory any more. Once theories tend to remain empirically unconfirmed for the whole range of a physicist's active career or longer, it becomes increasingly important to assess the theory's status already in its empirically unconfirmed state. In fact, being able to make such assessments becomes crucial for determining the relevance and status of the entire research field. The philosophy of science is confronted with the important questions on what grounds such assessments can be made and what status may be attributed to the corresponding theories. These questions become even more relevant – and at the same time more complicated – in a scientific context where empirical data is available but does not amount to conclusive evidence for or against a theory's viability. In those cases, the borders between unconfirmed and confirmed theories get increasingly blurred. Focussing on empirical confirmation under such circumstances appears artificial and rather unhelpful for grasping the overall context of theory assessment.

All this does not amount to a conclusive proof that a new philosophical analysis of theory assessment is actually required in the given situation. The fact that physics is not capable at this point of providing empirical confirmation for its theories may simply be taken as a serious and deplorable deficit of contemporary fundamental physics. It may seem questionable to reconsider the focus on empirical confirmation just because a certain scientific field under specific circumstances fails to meet its standards. After all, empirical confirmation has attained its crucial status in theory assessment for good reasons. It emerged as a core element of the scientific method that has brought about the stunning scientific and technological achievements which have shaped the world for the last few centuries. Rather than reconsidering the mechanisms of theory assessment, it might in this light be considered more adequate to acknowledge the crisis of a scientific discipline and wait for better times.

The account of so far unconfirmed theories that was given above offers significant arguments against that line of reasoning, however. The new philosophical questions which arise in the context of contemporary high energy physics do not just result from a longing for satisfactory strategies of theory assessment but result from the way contemporary high energy physics actually proceeds. We have seen that, even in the absence of empirical guidance, high energy physics today is characterized by a considerable directedness of theory building. The development of very specific and often highly predictive theories based on theoretical reasoning is striking. In a number of cases, specific conceptions have been established as solutions for specific conceptual or phenomenal problems without any alternatives in sight. Those conceptions play the role of influential and well-established theories. Though the lack of empirical confirmation is always seen as a regrettable obstacle, those unconfirmed theories are taken to be very trustworthy and reliable by their exponents.

All these characteristics of the research process in contemporary high energy physics suggest that theory assessment in the absence of empirical confirmation is in fact happening in the field and does play an important role in the evolution of fundamental physics today. The analysis of previous chapters demonstrated that the strategies involved in non-empirical theory assessment, while being weaker than strategies of empirical confirmation under "normal" circumstances, are rational and may become fairly convincing in specific contexts.

The characteristics of what one may call a dawning new phase of the scientific process in fundamental physics can be seen most clearly and most dramatically in string physics. The theory is possibly more difficult to confirm empirically than all other theories mentioned above. It is clearly the theory where finding a complete formulation requires the greatest conceptual and mathematical effort. It may also be the theory that is trusted the most by its exponents. For all those reasons, it made sense to try to develop the overall nature of non-empirical theory assessment in contemporary fundamental physics by looking at string theory. It is important to acknowledge, however, that the relevance of the questions addressed and of the solutions suggested transcends the case of string physics and deals with the status of fundamental physics in general. Even if string theory failed – and the assessment strategies which brought physicists to trust its relevance so strongly thus became more doubtful – the task of thinking about methods to assess theories in the absence of empirical confirmation would remain as urgent as ever.

5

Scientific underdetermination in physics and beyond

The previous chapters have analyzed the basis on which the importance of non-empirical theory assessment – and the assessment of limitations to scientific underdetermination in particular – has emerged as a consequence of the current dynamics in fundamental physics. To complete that picture, it is important to ask a slightly different question: to what extent are the developments so far discussed in the context of contemporary empirically unconfirmed fundamental physics rooted in earlier stages of physical development? Are they genuinely new or can they be understood as a continuation of tendencies which have characterized physical research already in the past? In the following, it will be argued that assessments of scientific underdetermination can be seen as an integral part of twentieth-century physics in two different senses. First, the increasing importance of non-empirical theory assessment is part of a more general development that characterizes the evolution of microphysics throughout the twentieth century. I will call that general development the marginalization of the phenomena. Second, a closer look at the mechanisms of physical reasoning reveals that the assessment of scientific underdetermination plays an unexpectedly important role also in empirically confirmed physics.

5.1 The marginalization of the phenomena

Throughout the twentieth century, one can identify a consistent tendency: the experimental signatures which confirm theoretical statements in fundamental micro-physics are moving towards the fringes of the phenomenal world and, at the same time, leaving increasingly wide spaces for entirely theoretical reasoning with little or no empirical interference. This, certainly, does not imply that empirical testing loses its crucial importance for scientific inquiry in high energy physics. The process is more subtle than that. It can be identified in a number of related but distinct contexts which shall be presented individually in the following.

Physics has encountered the development of what shall be called the marginalization of the phenomena in five different forms: (i) marginalization in an observational context; (ii) marginalization in connecting theory to experiment; (iii) marginalization in concept formation; (iv) marginalization in theory dynamics; and (v) the trust in theoretical conceptions is increasingly based on theoretical considerations. These are discussed below.

(i) The marginalization of the micro-physical phenomena in an observational context

The phenomena to be explained by topical microphysics have moved from the center of the observational realm towards remote and barely accessible areas of experimental testing. In the nineteenth century, micro-objects were introduced for explaining phenomenal laws like the laws of thermodynamics or basic features of chemistry. The phenomena guided by these laws, like e.g. air pressure or combustion, could be experienced by everyone looking at nature. When physicists started designing experiments capable of testing specific aspects of the microphysical theories, those experiments often produced subtle and minute phenomenological output, the observation of which required considerable technological support. The tested theories as a whole, however, still had very substantial consequences which could be observed by everyone. The first decades of the twentieth century saw microphysical theories being deployed in technical developments which changed and shaped everyone's life. From the utilization of X-rays to the building of nuclear bombs and power plants, from the development of television to the use of neutron physics in modern medical treatments, technological progress was and still is based on the fundamental theoretical concepts developed in the first half of the twentieth century.

In the second half of the twentieth century, the detachment of a theory's experimental consequences from everyday life reached a new level. Due to the short life-span of newly discovered particles, neither the particles nor the structure they disclosed had any immediate implications for the observable world which went beyond the precision experiments where they were found. The theories tested in these experiments have not acquired technological relevance until today and there are no indications that this could change in the foreseeable future

Signatures are limited to a few unusual lines on a set of pictures extracted from a particle detector. The specific structure of these lines remains entirely irrelevant if judged solely in the context of observable phenomena. They derive their significance solely from their theoretical interpretation.

From the 1970s onwards, finally, the quest for consistent unified theories of physical forces led to theories which predicted conceptual unifications whose characteristic length scales lay many orders of magnitude below the scales accessible at the time and thus added substantially to the motivation for increasing the experimental energy levels. Today, it requires the multi-billion-euro construction of a many kilometers-long particle collider and the sustained work of thousands of experimentalists to test physics at new energy scales. A large share of the experimental signatures of contemporary particle theories are, if at all, only accessible to one or two experiments in the world. The characteristic experimental signatures of advanced particle theories like grand unified theories or string theory lie entirely beyond the reach of any experiment imaginable today.

(ii) Marginalization of the phenomena in connecting theory to experiment

An increasingly complex theoretical machinery is required for connecting empirical signatures to specific theoretical claims. The patterns of particle traces observed on pictures extracted from particle detectors stand in a rather complex relation to the particles they are taken to confirm. Good examples of the complexities involved are the empirical identifications of those particles whose life-span is too short to leave a trace on a picture. The occurrence of such particles can only be inferred indirectly from the analysis of the vertices on the pictures extracted from the collider experiment. Vertices are "points" in those pictures where different particle traces meet. They are taken to represent locations where particles decay into other particles in the detector. The application of well-established theories of particle physics can lead to the conclusion that a certain frequency of specific kinds of vertices on the collected pictures can only be understood based on the assumption that some of the corresponding particle decays have happened via the creation of specific intermediate particles whose lifetime was too short to leave any visible traces themselves. The observation of a statistically significant number of vertices of the given kind is then interpreted as empirical confirmation of the specific particle in question. The analysis is of a statistical nature and does not allow identifying one specific individual vertex as being generated by the posited particle. Obviously, inferences of the described sort require a lot of fairly stable theoretical knowledge about particle physics. The more complex the theories and the corresponding experiments get, the longer becomes the "theoretical distance" from the empirical signature to the confirmation of a theoretical statement.

(iii) Marginalization of the phenomena in concept formation

The example of the macroscopic material object loses its power over scientific conceptualization. Early micro-physics took micro-physical objects to be small copies of the objects of the everyday world. By and large, they were supposed to share all of the latter's physical properties. In the late nineteenth century, doubts grew whether it was possible to reconcile this naïve conceptual approach with the observed phenomenology. The lack of precision measurements of micro-physical objects did not allow the development of any specific and empirically successful alternative concepts, however. The discovery of specific inconsistencies of a classical understanding of the atom as it emerged based on the experiments of the early twentieth century finally led to quantum mechanics and the abandonment of the classical paradigm. Quantum mechanics dropped the classical scientific understanding that an object must have a precise location and momentum and must move according to deterministic laws. Furthermore, it led to a conception of particles that ran counter to a classical understanding of particle individuation. A little earlier, a similar distancing from intuition had occurred in the understanding of space and time based on the counter-intuitive conceptions of special and general relativity. The grand revolutions of relativity and quantum mechanics were followed by other infringements on intuition. Quantum field theory gave up on the assumption that a precise number of objects must exist at each point in time. The standard model of particle physics abandoned the understanding of mass as a primitive characteristic of objects and explained it in terms of a specific interaction structure. Concepts of quantum gravity may eventually imply giving up the fundamental character of the concepts of space and time. All these developments are based on mathematical reasoning aimed at providing a theoretical structure that is internally consistent and fits the empirical data. While the observed phenomena started out in a double function as empirical testing criterion and a raw model for the posits of micro-objects, today the second function has been all but dissolved.

(iv) Marginalization of the phenomena in theory dynamics

In contemporary particle physics, theory has replaced experiment as the primary driving force behind scientific progress. In earlier microphysics, experiments typically guided the evolution of scientific theories by providing new discoveries which demanded new theoretical explanations. The structure of atoms and nuclei, the existence of a large number of new particles, the different types of nuclear interactions and phenomena like CP-violation were first discovered by experiment and later described theoretically. The recent evolution of fundamental micro-physics altered this balance between theory and experiment. Over the last four decades, particle physicists have become accustomed to a scientific environment where theory building is guided mainly by the theoretical insufficiencies of the existing theories. New theoretical schemes mostly are not introduced due to empirical refutations of their predecessors but rather in order to solve consistency problems of existing theories[1] or to explain theoretical features of existing theories that look unnatural in the old framework.[2] The resulting new theories predict new empirical phenomena which are then sought to be tested experimentally. The prediction and later discovery of the standard model particles followed this pattern; the experimental discovery of supersymmetry is a current experimental goal that falls into the same category. Experiment thus is largely (though not entirely, certainly[3]) reduced to the role of a testing device for already existing particle theories. In addition, as already emphasized, the time lag between theoretical posits and their empirical testing is steadily increasing. While experiments in the 1970s mostly tested theoretical predictions which had been made just a few years before, the current experiments at CERN's Large Hadron Collider search for particles whose existence has been proposed in the 1960s (in the case of the Higgs Boson) or around 1980 (in the case of the supersymmetric particle spectrum).

The final point addresses the phenomenon that was discussed in the first part of this book.

(v) The trust in theoretical conceptions is increasingly based on theoretical considerations

It would be an exaggeration to speak of a genuine marginalization of the phenomena in this case. Empirical testing of theoretical claims remains the crucial

---

[1] Examples are the particle physics standard model or string theory. The standard model achieved the renormalizability of strong and weak interactions; string theory is an attempt to solve the problem of infinities which arise when gravitation is unified with point-like particle physics.

[2] Examples are grand unified theories and low energy supersymmetry. Grand unified theories aim at explaining the fact that the three microphysical coupling constants have (approximately) the same value at a certain energy scale; low energy supersymmetry offers an explanation for the huge scale difference between the electroweak scale and the GUT respectively Planck scale.

[3] Not all testable particle properties are predicted univocally by particle theory. The measurement of neutrino masses, to give one example, was not implied by the standard model of particle physics (though it appeared fairly natural in that framework). In cosmology, the measurements of a cosmological constant were a genuine surprise.

final goal in contemporary particle physics just like in other scientific fields. After all, large sums are invested in the construction of particle colliders in order to keep particle theory empirically testable. However, by infringing on the last strong bastion of empirical dominance, the growing importance of non-empirical theory assessment contributes to the overall shift of the balance between theory and experiment and therefore can be seen as part of the overall process that we have called the marginalization of the phenomena.

The increasing importance of non-empirical theory assessment thus can be understood as part of a broader development towards an increasing role of theory within the scientific process in physics. Up to this point, it may seem that the significance of limitations to scientific underdetermination only becomes conspicuous at the last stage of this development. In the following, it will be argued, however, that it actually plays a crucial role for understanding the overall structure of the marginalization of the phenomena. To explain this point, we need to have a closer look at the five kinds of marginalization of the phenomena. It turns out that an interesting tension between different kinds can be found, which may be best described by relying on the concept of scientific underdetermination.

Among the various kinds of marginalization of the phenomena, one can identify two different tendencies which pull in opposite directions. On the one hand, there are those kinds which provide additional degrees of freedom for theory construction and thereby potentially contribute to scientific underdetermination. Since successful predictions of novel phenomena have been shown to be directly linked to limitations to scientific underdetermination, those kinds of marginalization of the phenomena constitute a potential threat to the predictive power of microphysics. Two of the five kinds listed above fall into this category. The demise of naïve ontology stated in point (iii) removes the corresponding constraints on theory building. Once the constraints posed by the requirement that theories are largely consistent with a naïve ontology of microphysics have vanished, all mathematically coherent ways of structuring the observed phenomena must be taken to be potentially physically viable. The resulting widened spectrum of all mathematically coherent structures might well be assumed to be sufficiently large for allowing the reproduction of all imaginable sets of future empirical data. As long as there are no a priori reasons for taking any of those structures to be more likely than any other, such a scenario would lack a plausible perspective of successful predictions.

A second development that works in the same direction is described in point (ii): an increasing amount of theoretical reasoning is required in contemporary particle physics for connecting the empirical data to the scientific object it is supposed to confirm. This increasing conceptual complexity might be expected to raise the chances of finding alternative theoretical structures which construe the connection between empirical data and theoretical implications in different ways. Such alternatives, in turn, may imply different empirical predictions with respect to novel phenomena and therefore can endanger predictive power.

Given those two developments, the high predictive power of particle theories with respect to novel phenomena in the 1970s and 1980s seems even more remarkable. While the predictive success of those theories appears to be at variance with types (ii) and (iii) of the marginalization of the phenomena, however, it is actually reflected by two other kinds of marginalization which indicate stronger limitations to scientific underdetermination.

The leading role of theorizing in the dynamics of particle physics (point (iv)) is based on the fact that consistency arguments often seem to force theoretical development to move in a certain direction even in the absence of empirical confirmation. This implies that no alternative theoretical options seem available, which may indicate limitations to scientific underdetermination. Point (v), the increasing trust in non-empirical theory assessment, reveals that scientists do rely on assessment of limitations to scientific underdetermination.

Relating points (iv) and (v) to points (ii) and (iii) highlights the important message regarding the question of limitations to scientific underdetermination that can be drawn from the overall phenomenon of the marginalization of the phenomena. It is a highly non-trivial observation that, despite the demise of the role of naïve ontological constraints and the growing theoretical distance between experimental signature and fundamental theory, scientific underdetermination reveals strong limitations in the context of fundamental physics. The fact that such limitations nevertheless occur may be taken to constitute one of the most important general results of contemporary fundamental physics.

Taking this message seriously leads to the conclusion that particularly strict limitations to scientific underdetermination are at work not only within the current context of scientific theories which have not found empirical confirmation for long periods of time but already at earlier stages of microphysics where empirical confirmation was available. In fact, the empirical confirmation achieved in such contexts is of crucial importance for identifying limitations to scientific underdetermination in the first place. The question thus arises whether scientific reasoning at earlier stages already did take into account that element of theory assessment. The following sections will be devoted to that question.

5.2 The question of microphysical objects

It has been emphasized already in Chapter 1 that assessments of scientific underdetermination always played an informal role as a basis for assessing the chances of success for empirically unconfirmed theories. Individual scientists at times put a lot of emphasis on non-empirical methods of theory assessment that can be framed in terms of implicit assessments of limitations to scientific underdetermination.

The best-known example in case is Einstein's trust in his theory of general relativity. After the theory had been confirmed empirically by Eddington's measurement of starlight bending in 1919, Einstein famously remarked that he had been fully convinced of the theory's validity all along: general relativity just was too beautiful to be false. Translating Einstein's remark into more specific terms, he was conveying that (1) he considered it highly unlikely that there existed more than one theory that was in agreement with the data available at the time he completed his theory of general relativity and was as beautiful as general relativity and (2) he considered it highly unlikely as well that, if such a beautiful theory existed, a less beautiful one would be correct and the beautiful one would be false. Step (1) of the argument, reconstructed in this way, represents a straightforward example of an assessment of scientific underdetermination with regard to scientific theories which satisfy the additional criterion of beauty.

Despite strong cases of the above kind, however, the canonical paradigm of theory assessment understands such considerations as subjective assessments which do not generate genuine scientific knowledge about the external world. Earlier stages of this book have suggested that this canonical paradigm of theory assessment is seriously at odds with the way empirically unconfirmed theories and claims are actually assessed in contemporary fundamental physics. A little later, it will be demonstrated that this is true not just with respect to speculative theories like string theory but also within the context of well-established standard model physics.

I want to pursue a more far-reaching approach, however. I will argue that assessments of scientific underdetermination play a crucial and systematic role already in a context where one would not expect it at first glance: they are built into the overall understanding of what it means to acquire empirical confirmation of theoretical scientific claims. I will argue for this claim by discussing two specific contexts. In the present section, I will have a look at the emergence and confirmation of empirically supported atomism in the later part of the nineteenth and the first decade of the twentieth century. Section 5.2 will then analyze present-day empirically confirmed high energy physics. Following that analysis, I will return to cases of empirically unconfirmed theories and will demonstrate that the roles played by assessments of scientific underdetermination in the two different contexts are remarkably similar.

Atoms and molecules were conjectured for long periods of time without any strategy of direct empirical testing. Rooted in the philosophical atomism of antiquity, the belief in the existence of atoms was widespread among the physicists and chemists of the seventeenth and eighteenth centuries. In the nineteenth century, substantial improvements in atomist conceptualization as well as experimental technology led to an increasing body of what was at the time understood by many as empirical support for atomism. Supportive data for atomism came from chemistry as well as from physics. In chemistry, Dalton's law of constant proportions between the weights of elements in chemical substances found a plausible interpretation in terms of atoms which formed compounds (molecules). The discovery of isomers, chemical substances which consisted of the same elements to the same proportions but had different chemical properties, provided an even stronger argument for atomism. As first suggested by Pasteur and later exemplified in a specific model for methane by Le Bel and van t'Hoff, isomers could be understood in terms of molecules consisting of the same atoms but arranged in different spatial structures. The fact that no satisfactory alternative explanation of isomers was forthcoming suggested, in the eyes of many observers, that the viability of atomism could be inferred based on abductive reasoning. In physics, one important argument for atomism was the possibility to reduce thermodynamics in a coherent way to a microphysical interpretation. Work by Daniel Bernoulli, James Clerk Maxwell, Ludwig Boltzmann and others demonstrated that such a construction was quite accurate and capable of explaining a wide range of physical properties, from the second law of thermodynamics to quantitative properties of the viscosity of gases and the calculation of the specific heat of gases. (In the latter context, atomism did face some problems that weren't solvable before the advent of quantum mechanics but still was able to reproduce important features correctly.) The capacity of atomism for offering explanations of a number of different empirical phenomena led many scientists to have considerable trust in the concept's validity. The posit of the atom thus can be taken as one strong example of a prediction whose viability was believed based on empirical observation in connection with theoretical background reasoning. Nevertheless, no one at the time was ready to call atomism an empirically confirmed theory. After all, no direct observational contact had been made with atoms.

The strong trust in atomism despite the general consensus that atoms could not be called empirically confirmed is reminiscent of the situation of string physics and other theories today. It therefore seems instructive to ask whether the situation at the time led towards a reconsideration of the scientific criteria for scientific knowledge along the lines suggested in Part I of this book. More specifically, did the debate on atomism address questions of scientific under-determination? We will see that attempts to redefine theory assessment as well as allusions to the question of underdetermination can indeed be found in the atomism debate. However, at least in the most conspicuous cases, they did not occur together.

Of particular interest with respect to the question of the status of a scientific theory is an analysis carried out by James Clerk Maxwell in the early 1870s. Maxwell believed that the multitude of indications for the viability of atomism was so strong that it was inadequate to call atomism a speculative hypothesis any longer. Atomism in his view seemed likely to constitute knowledge about the external world. However, Maxwell fully shared the general understanding at the time that the empirical support for atomism did not amount to empirical confirmation. Moreover, he understood well that the canonical paradigm of theory confirmation did not leave any middle ground between an empirically confirmed theory, on the one hand, and a hypothetical speculation on the other. The situation was particularly difficult because, at the time of Maxwell's writing, it was by no means clear whether and on what grounds actual empirical confirmation of atoms could ever be achieved. Empirical confirmation seemed conceptually bound to observable entities. All empirical data in support of atomism thus had to fall short of empirical confirmation as a matter of principle. Given such an understanding, the problem was not how to assess the status of atoms before they had found empirical confirmation. It rather was how to assess the status atoms could obtain at all.

Maxwell dealt with this problem by defining an intermediate status between being empirically confirmed and being a mere hypothesis. A theory that had obtained that intermediate status according to Maxwell did represent probable scientific knowledge. Maxwell's considerations have recently been analyzed in a very interesting paper by Peter Achinstein (2010). The details of Maxwell's construction are of no crucial importance to our discussion. What is important, however, is the fact that the situation of atomism from the 1860s until the emerging consensus regarding the empirical confirmation of the atom around 1910 did show some resemblance with the status of high energy physics today and, for a while, opened up a perspective towards an extension of the conception of scientific knowledge beyond the limits of empirical confirmation. The path eventually chosen by physicists, however, led back towards the monopoly of empirical confirmation when atoms eventually were taken to have acquired the status of an empirically confirmed concept.

Maxwell's analysis did not focus on the question of limitations to scientific underdetermination. Some anti-atomist positions did, however. Despite strong corroborative support for atoms, anti-atomism was widespread in the second half of the nineteenth century. Leading scientists like the chemists Jean Baptiste Dumas, Hermann Kolbe and Wilhelm Ostwald or the physicists Ernst Mach and Pierre Duhem were outspoken opponents of atomism. Their positions were motivated by various considerations. One pivotal argument hinged on the allegedly naïve ontological intuition behind the concept of the atom which, in the eyes of many anti-atomists, did not square well with a modern scientific outlook that was expected to focus on abstract mathematical reasoning. Pierre Duhem (1954) connected this argument to the question of underdetermination. He rejected atomism because he believed that the choice of a specific ontology was fundamentally underdetermined by empirical data. According to his argument, all ontological commitments, whatever their nature, had to be treated in a highly flexible way in order to keep them adaptable to the given empirical data. If such adaptations were allowed for, however, any ontology could be adapted to any empirical data pattern with appropriate measures. Thus, atomism just like any other ontological preconception could be made compatible with any observed phenomenology. Atomism thus did not constitute a meaningful scientific posit at all because it did not imply specific empirical predictions. Duhem's argument can be read as relying on a version of the principle of underdetermination. The focus of Duhem's reasoning, however, is on ontology rather than on prediction. While the basic question of scientific underdetermination as it has been defined in this book is concerned with possible alternative scientific theories which are in agreement with the present data but offer different predictions with regard to future experiments, Duhem emphasizes the possibility of alternative theories which are based on a different ontology.

The eventual rejection of Duhem's position by the physics community after the general acceptance of atomism as an empirically confirmed theory actually treated the question of underdetermination in a way that came closer to my treatment of scientific underdetermination than Duhem's position itself. One important precondition for that acceptance was the fact that no non-atomist theory had been found that could compare to atomism with respect to its success in accounting for the wide range of new empirical phenomena. The denser that web of empirical data became, the more unlikely it seemed that any non-atomist theory could be as successful as atomism in this respect. The shift from the understanding that atomism was supported by empirical evidence but could not be called empirically confirmed to the understanding that atomism was indeed an empirically confirmed theory was a gradual one. Many elements which eventually were taken to be crucial for speaking of empirical confirmation had been present in a weaker form already at earlier stages of atomic research. Already in the later parts of the nineteenth century, atomism had been able to make correct predictions, offer a coherent and unified explanation of many seemingly independent phenomena and was generating fruitful new theoretical and empirical research. Still, there remained the general sentiment, characterized above, that observation had to be immediate to amount to empirical confirmation.

This sentiment in fact made a lot of sense based on considerations about scientific underdetermination. If the Newtonian theory of gravity predicts a planet's path and the planet can then be observed moving along that path, one can talk about having confirmed the theory because the ontological state that is in accordance with the theory's core prediction has been directly observed. If the atomic theory predicts certain corrections to the laws of thermodynamics and those predicted deviations are then measured, the ontological setup implied by the theory has not been observed. Thus, a genuine worry about underdetermination arises. Another theory might offer the same prediction without positing atoms. It may therefore be justified to use the theory to the extent it saves the phenomena but it seems questionable to speak of an empirical confirmation of atoms.

The same fundamental worry remained valid in principle after 1910 – and indeed up to this day. It provides the basis for anti-realist positions in the philosophy of science. Bas van Fraassen's main argument against scientific realism, to give a prominent example, is based on an argument of underdetermination of theory building by the empirical data. While the philosophical question thus was here to stay, the scientists' attitude towards empirical confirmation changed after 1910. From that time onwards, it was considered legitimate to talk of the empirical confirmation of unobservable objects.

The honor of constituting the first genuine experimental confirmation of atoms is usually awarded to the experiments carried out by Perrin between 1908 and 1911. Perrin was able to generate very similar values for the Avogadro constant (the number of molecules per mole) based on a number of different experimental methods and theoretical ways of reasoning. He thus offered a web of specific quantitative predictions regarding observable quantities based on the atomic hypothesis. The reasoning that led to acknowledging atoms on that basis may be understood as a kind of "no miracles" argument: it would be inexplicable why those various predictions turned out correct if the basic atomistic idea behind their calculation was entirely wrong. A number of pragmatic arguments were playing a role. One of them was the argument of no alternatives: it became increasingly hopeless to look for non-atomist theories which were as powerful as the atomistic ones. Another one was the obvious conduciveness of atomism to scientific progress. Atomism created new directions for research which proved fruitful on a regular basis. Other arguments had a more immediate intuitive basis: experiments like Rutherford's scintillation detector for the first time allowed relating individual observational events causally to individual microphysical objects.

All these arguments did not remove the basic threat of possible unconceived and non-atomist alternatives to the present theories.[4] Jointly, they created a situation, however, where it seemed reasonable to disregard that possibility and interpret the data univocally in terms of the atomic theory. It just appeared increasingly unlikely that any viable non-atomistic alternative existed. This step of disregarding scientific underdetermination set the example for all ensuing empirical confirmations of microphysical objects. The assessment of limitations to underdetermination that was required for accepting the statement that atoms constituted knowledge about the world was thus absorbed into the new concept of empirical confirmation of unobservable objects. Rather than establishing a new status for well-established unobservable objects – as Maxwell had proposed in the 1870s – the physics community was led by the overwhelming evidence for atoms after 1910 to equate the epistemic status of atoms and other microphysical objects to the status of observable objects. Both could be called empirically confirmed if certain conditions were met.

The decision of 1910 solidly re-established the dichotomy between empirical confirmation and hypothetical speculation based on a modified concept of empirical confirmation. Assessments of limitations to scientific underdetermination now were happening on both sides of the divide. Those related to the trust in unconfirmed theories – as exemplified by Einstein's early trust in his theory of general relativity – were considered individual assessments which did not carry inter-subjective weight and therefore did not contribute to scientific knowledge. Those related to empirical confirmation of scientific objects, to the contrary, were hidden within the latter concept, which created the canonical perspective that allowed attributing the status of scientific knowledge to microphysical claims.

5.3 Empirical confirmation in high energy physics

The strength and stability of the support given to empirical confirmation by implicit assessments of limitations to scientific underdetermination can be best seen in contemporary advanced contexts where intuition-based ontology has all

---

[4] Roush (2006) argues that Perrin's method of measuring Brownian motion in fact excludes unconceived alternatives and therefore justifies a realist understanding of atoms. I fully concur with Stanford (2009), however, who points out that Roush's analysis is based on taking for granted a specific conceptual framework that could itself be questioned. For a more extensive discussion of Roush's approach, see Section 7.2.

but vanished and the theoretical distance between empirical signatures and the fundamental theory has become very large indeed. I want to exemplify this by having a look at contemporary empirically testable high energy physics, more specifically at the case of the empirical search for the Higgs particle.

As described in Chapter 4, until 2012 the Higgs particle represented the last remaining unconfirmed prediction of the standard model of particle physics and was searched for during the LHC experiments. Theoretical arguments suggested that Higgs particles should be produced there in significant numbers when protons were smashed together with the energies available. Due to its high mass, the Higgs particle was expected to decay too fast to leave a visible trace or gap[5] on decay pictures taken in detectors. Thus, the particle had to be identified on those pictures by looking at vertices. As described in Chapter 4, the theoretical understanding of particle physics can lead towards interpreting certain decay patterns in terms of particle decays via specific kinds of particles within specific vertices visible on the pictures. The collection of a statistically significant number of clearly identified candidates for the occurrence of a Higgs particle eventually was acknowledged as an empirical confirmation of the Higgs particle.

Confronted with a set of experimental signatures which count as candidates for Higgs particles, it was crucial to understand whether all potential explanations of the set of signatures in question which did NOT rely on the existence of a Higgs particle could be confidently ruled out. A number of potential alternative explanations had to be checked. It was known that Higgs-like individual signatures could be caused by exponents of other particle decays. It thus had to be checked whether the measured number of Higgs-like events lay significantly above what would have been explicable based on the totality of alternative explanations. Beyond this consideration of all known alternative explanations of individual events within the body of theoretical physics, other eventualities had to be taken into account. There could have been all sorts of technical errors related to the involved experimental devices. There also might have been systematic errors in the analysis of the data. Finally, the signatures could also have been produced by an entirely new physical phenomenon that was not accounted for by the existing particle theories at all. Assessing all possible sources of error constitutes a crucial and fully acknowledged part of particle physics. It relies on strategies for looking at those alternatives which can be

---

[5] The standard model Higgs particle, being electrically neutral, could only produce a gap on a decay picture, between the point where it is produced and the point where it decays. Supersymmetric models predict charged Higgs particles, however, which could in principle leave a trace, if it weren't for their short lifetime.

checked as well as on providing reasons for attributing a low probability to the possibility that any other theoretical explanation which has not been thought of applies. The process thus provides a conspicuous example of a situation where assessments of limitations to underdetermination directly enter the scientific process.

One aspect of this assessment is of particular interest for our discussion. As mentioned above, before accepting a set of signatures as confirmation for the Higgs particle, one had to be confident that the signatures are unlikely to be explicable by some entirely new physical phenomenon for which no theoretical ideas had been developed so far. Naturally, the possibility of such a new phenomenon can never be excluded analytically. The only basis for discarding it relies on a meta-inductive argument. Given that there have not been all that many entirely new types of phenomena in the past, it would seem a strange coincidence if such a new phenomenon produced exactly those signatures which are taken to confirm the theory in question (in our case the particle physics standard model that predicts the Higgs particle, or maybe one of its extensions) in an independent way. There remains the possibility, of course, that an entirely new and fundamentally different theory explains the same new experimental signatures as the theory to be tested due to some deep structural connection between the two theories.[6] Dealing with this worry, the theorist can do no more than stress that she does not see any indications for such a scenario. The step towards accepting some signature as confirmation for the Higgs particle therefore relies crucially on an inference from a statement on the scientist's intellectual horizon (namely that the scientist cannot think of any realistic alternative scenario that would lead to the empirical data taken as confirmation of the Higgs particle) to a probability statement about the world (namely that most likely there ARE no other conditions in the world which can create that empirical data). In other words, the scientist carries out an assessment of limitations to scientific underdetermination.

A strictly empiricist reader might, at this point, put forward the following worry. From an empiricist point of view, empirical confirmation of a theory by definition just constitutes successful empirical testing of its predictions. Inferences to a theory's truth or a particle's real existence are not part of the scientific process for the very reason that assessments of scientific underdetermination cannot establish scientific knowledge. The empiricist thus cannot accept the need for an inference of the kind stated above as long as that inference simply leads to the acceptance of an ontological posit.

[6] We have encountered this possibility already in our discussion of arguments for limitations to scientific underdetermination in Section 3.1.

An answer to this worry can be based on the distinction between two different specifications of the empiricist tenet. A "milder" version of empiricism holds that theory confirmation does not establish the (approximate) truth of the confirmed theory. The discussion in this chapter has no quarrel with this understanding. It does not imply any conclusions regarding the reality of empirically confirmed theories. A more radical empiricist claim denies that empirical confirmation can establish anything beyond the fact that the theory has passed the empirical test in question. This radical claim is being rejected in the present discussion. As argued at several stages in this book, the trust in a theory's future predictions based on the theory's past empirical confirmations constitutes an essential part of scientific rationale and cannot be denied without distorting the scientists' attitude towards their confirmed theories. The trust in the predictions of empirically confirmed theories thus constitutes an essential implication of theory confirmation. It is this latter aspect rather than the question of the reality of confirmed scientific objects that is addressed in the present argument. It is claimed that, provided that scientists speak of the empirical confirmation of a theory in order to express their trust in all predictions implied by that theory in the given regime, they cannot establish empirical confirmation without heavily relying on assessments of scientific underdetermination.

5.4 Non-empirical assessment of the Higgs theory

Having thus established the significance of assessments of scientific underdetermination in the context of theory confirmation, I now want to return to the case of the assessment of scientifically unconfirmed theories. Remaining within the context of Higgs physics, I will argue that the rationale behind the understanding of empirical confirmation and the rationale behind non-empirical theory assessment are indeed very closely related.

While the previous section was devoted to analyzing the confirmation of the Higgs particle at the LHC, I now want to consider the situation before 2012, when the Higgs particle had not yet found empirical confirmation. At the time, it was conjectured for entirely theoretical reasons. The Higgs mechanism, i.e. the theoretical structure which implied the existence of the Higgs particle, had to be included in the standard model in order to allow for the observed masses of matter particles and for massive electroweak gauge bosons. The standard model just could not account for the occurrence of massive objects in the observed world if the Higgs sector was simply left out. Thus, the Higgs mechanism was an essential part of the standard model and did not constitute an independent new theory in its own right like, e.g., string theory. Nevertheless, it was based on a separable set of theoretical posite and its predictions could be distinguished from those based on other segments or principles of the standard model: it predicted at least one new particle (the Higgs particle)[7] and a certain structure of that particle's interaction with itself and with matter. The Higgs sector therefore constituted a separable "module" of the standard model, a kind of sub-theory whose viability could be discussed separately from the other parts of the standard model.

Despite its unconfirmed status, the existence of the Higgs particle was rarely doubted by particle physicists over the last three decades. The particularly strong trust in its existence before empirical confirmation makes the Higgs particle an interesting example of the power of non-empirical theory assessment in high energy physics. The path of reasoning that led to the Higgs particle's prediction involved a number of separate steps which started at the level of fairly general considerations. First, there was the physicists' understanding that the description of relativistic phenomena in microphysics required quantum field theory. Within quantum field theory, a fully consistent calculation of scattering processes was understood to require the theory's renormalizability. Renormalizable interacting quantum field theories, in turn, were understood to require a gauge symmetric structure. A gauge symmetric structure was understood to be compatible with massive interaction particles and a spectrum of different fermionic masses only if the corresponding gauge symmetry was spontaneously broken. Coherent spontaneous symmetry breaking, finally, required the existence of a Higgs Boson.

The question arises to what extent was it justified to believe in the prediction of the Higgs particle based on the given rather complex sequence of theoretical arguments? The line of reasoning sketched above could have been doubted on two grounds: first, the possibility could have been considered that no consistent physical description of the phenomena in question existed at all; second, it could have been suspected that physicists had overlooked theoretically viable alternatives which would have broken one of the stated inferences. The first worry, though by no means inherently absurd, goes counter to today's deeply entrenched convictions regarding the range of scientific reasoning. The second one, however, is the standard worry of scientific underdetermination. When the Higgs particle was first predicted, the first steps of the line of reasoning given above seemed quite uncertain. The possibility seemed realistic that the entire framework of quantum field theory was inadequate for offering a fully coherent description of the nuclear forces;[8] and even if one had accepted the framework of quantum field theory at the given energy scale, the deployment of gauge symmetries could have

---

[7] Which may be either fundamental or a constituent particle.

[8] It should be noted that an argument of this type in fact does apply once one takes the additional step to look for a joint description of microscopic forces and gravitation. Gauge field theory is not able to provide such a joint description, which leads the way towards string theory.

been questioned by suspecting that another equally viable solution to the problem of renormalizability had simply been overlooked. These worries prevented excessive trust in the standard model before it had been tested empirically. Both worries seemed largely defeated in recent decades, however, first due to an improved conceptual understanding of the situation and second since experiments had provided ample confirmation for the validity of gauge field theory at the electroweak scale.

The second part of the line of reasoning sketched above, the steps from the assumption of the validity of gauge field theory to the prediction of the Higgs particle, seemed to be rather solid. On the basis of gauge field theory, the observed mass spectrum of elementary particles enforced the spontaneous breaking of electroweak gauge symmetry based on straightforward mathematical reasoning. The final step towards the Higgs field was also well understood and there seemed to be no really satisfactory way to avoid it.[9] Physicists in recent decades therefore believed quite strongly in the existence of the Higgs particle despite the lack of empirical evidence. Nevertheless, purely theoretical reasoning could not remove all doubts. It could not be fully excluded that the overall scheme of gauge field theory and the standard model, though having been confirmed in so many cases in the past, had to be replaced by a completely different theory in order to describe the generation of masses correctly. Particle physicists had two arguments why they considered that scenario unlikely. First, it would have been unclear why the standard model had been so successful in experiments at the electroweak scale in the past if it turned out to be the wrong theory at that energy scale after all. Second, there were no indications at hand as to what a convincing alternative theory could look like.

In the end, the question, once again, boiled down to an assessment of the limitations to scientific underdetermination: how plausible did it seem that an alternative theory was capable of reproducing all the data that had been taken as confirmation of the standard model in an equally satisfactory way without

---

[9] Physicists still have different opinions on the question whether the Higgs is an elementary or a composite particle and whether one or several Higgs particles should be expected. The basic principle that at least one scalar particle had to exist in order to make gauge field theory work, however, was considered fairly trustworthy. Still, it should be noted that the question of possible alternatives to the Higgs mechanism also constitutes an example for the volatility of no alternatives arguments. While it was for a long time believed that no alternatives could be found at all, more recent analysis has led to the emergence of so-called Higgsless models which aimed at explaining mass terms based on brane physics and duality principles without the introduction of Higgs particles. Since the consistency of those models seemed questionable and they were not considered overly attractive on various accounts, they never weakened the trust in the Higgs particle to any substantial degree. Still, theoretical reasoning for the existence of the Higgs particle in recent years may be better described in terms of an inference to the best explanation rather than in terms of an inference to the only discernible candidate for an explanation. Be this as it may, the crucial role of assessments of scientific underdetermination remains clear.

enforcing the existence of a Higgs field? Particle physicists largely shared the understanding that this scenario seemed rather unlikely. The final word, though, had to be spoken by experiment.

The non-empirical reasoning for the Higgs theory's viability was focussed on NAA and MIA as arguments for limitations to scientific underdetermination. That particular focus was related to the specific status of the Higgs theory as an essential "module" of a larger, empirically confirmed theory. The argument of no alternatives played a crucial role and was particularly effective due to the high degree of empirical confirmation of the standard model. Based on that confirmation, gauge field theory looked sufficiently well-established for accepting it as a viable framework at a basic level of analysis. The assessment that no satisfactory alternatives to the Higgs theory seemed possible within the context of gauge field theory thus was taken to carry considerable weight. The fact that UEA, the argument of surprising explanatory interconnections, did not play an important role is related to the specific position of the Higgs theory as well: since the Higgs theory constitutes a very specific construction within another theory, its potential to interfere with other physical contexts is more limited than in the case of a fully independent and autarkic theory. MIA, the argument of empirical success in related theories, of course played a particularly important role because it could rely on the history of empirical confirmation of the Higgs theory's embedding theory, the standard model.

5.5 The confirmation of the top quark

Let us now move one more step away from the case of unconfirmed theories. The standard model made a large number of empirical predictions which were later confirmed by experiment. Concluding a series of predictive successes, the confirmation of W- and Z-Bosons in 1983/84 was widely understood as a decisive step towards a conclusive establishment of the standard model as a viable description of physics below the electroweak scale. After the confirmation of W- and Z-bosons, it took ten more years until the next predicted standard model particle, the top quark, could be confirmed by experiment. During that decade, no one seriously doubted that the top quark would eventually be found. Unlike in the case of the Higgs particle, the prediction of the top quark was not based on any new theoretical concept. It relied on the same standard model principles which had already been tested and empirically confirmed several times. Believing in the existence of the top quark thus could be understood straightforwardly in terms of having trust in the predictions of a well-confirmed and well-established theory.

Nevertheless, the top quark constituted a new phenomenon, a new particle whose existence could not be taken for granted based on simple inductive inference. Thus, belief in the top quark's existence could be questioned based on the principle of scientific underdetermination. In principle, it might have been the case that another theory existed, which was empirically equivalent to the standard model up to the energy scales testable in the mid 1980s but which did not imply the existence of the top quark. If any such theory had been discovered in the late 1980s, trust in the top quark would have been reduced significantly. Vice versa, the trust physicists did have in the top quark's existence was based on the assessment that such an alternative was unlikely to exist. In other words, physicists relied on NAA, the argument of no alternatives, just like they do when assessing the chances for the viability of a fully fledged unconfirmed theory. Furthermore, their trust was bolstered by MIA type reasoning based on the standard model's past predictive success. An important point is nicely illustrated by this situation. One finds a striking similarity between empirical confirmation of the theory itself, on the one hand, and corroboration of one theory by the confirmation of other theories, on the other. Looking at specific so far unconfirmed predictions, the two kinds of argument play a very similar role. Scientists believed in the top quark before 1994 based on the previous predictive success of the standard model. In a similar way, the very same successes of the standard model today bring them to believe – to a certain degree – in the predictions of low energy super-symmetry or in the viability of string physics.

The notion of limitations to scientific underdetermination provides a reason for this continuity. The question to be asked when assessing the chances for some specific prediction being viable is the same in both cases: is it reasonable to assume that scientific underdetermination based on the available empirical data is limited to an extent that alternative theories which are also compatible with the known data but make different predictions are unlikely to exist? In both cases, an answer to this question is attempted based on a meta-inductive argu-ment: past predictive success demonstrates at an empirical level that limitations to underdetermination are in fact important in the given context. Given that other considerations (i.e. NAA and UEA) suggest that the present context is similar to those where empirical success did occur, it seems legitimate to infer a likelihood of empirical success also in the given case. If the empirical successes were achieved by the very same theory, the case for a sufficient similarity between those situations and the present one looks most convincing, of course. The difference to the case of other comparable theories is gradual, however. The character of the argument is the same and the same types of risk have to be taken in both cases.

We thus observe a continuity that leads from the evaluation of unconfirmed predictions of novel phenomena by empirically well-confirmed theories to the case of unconfirmed "modules" of confirmed theories, like the Higgs mechanism, and eventually to non-empirical theory assessment in cases of entirely unconfirmed theories like string theory. In all those scenarios the same basic principles of non-empirical theory assessments are deployed.

We have now identified assessments of scientific underdetermination in two different contexts of analysis. The classical paradigm of theory assessment takes them to carry fundamentally different epistemic weight depending on the context. The analysis of sources of error in the interpretation of empirical data and the assessment of possible alternatives to empirically confirmed hypotheses, as addressed in Section 5.2, are acknowledged by the canonical paradigm as an integral part of the acquisition of scientific knowledge. Assessments of scientific underdetermination thus play a crucial and scientifically fully respectable role in the process that leads towards the recognition of experimental signatures as empirical confirmation of a scientific hypothesis. They provide the basis for acknowledging the existence of a certain kind of particle or the viability of some abstract scientific principle like spontaneously broken gauge symmetry.

The classical paradigm of theory assessment denies the status of scientific knowledge to non-empirical theory assessment as described in Sections 5.3 and 5.4. Non-empirical theory assessment is merely acknowledged to be helpful in providing some idea of a theory's chances of scientific success. The latter must then be achieved based on empirical confirmation. The presented close structural similarities between the roles of assessments of scientific underdetermination in both contexts raise serious doubts whether a fundamental epistemic distinction between the two contexts of assessments of limitations to scientific underdetermination can be coherent at all.

Denying the status of scientific knowledge about nature to *all* empirically unconfirmed scientific claims would have to be based on the assertion that assessments of limitations to scientific underdetermination are incapable *in principle* of providing a basis for scientific knowledge about nature. As the example of the Higgs particle shows, however, assessments of limitations to scientific underdetermination of the very same kind used for arguing in favor of empirically unconfirmed theories enter the analysis of the empirical data itself. In both cases, there is the threat of potential new theoretical schemes which might offer interpretations/solutions for the given experimental signatures/ theoretical problems which have not been thought of so far. And both kinds of reasoning crucially rely on offering theoretical reasons for confidently taking the existence of such alternatives to be unlikely. The notion that the application

of theoretical assessments of scientific underdetermination as it is applied to
assessing empirically unconfirmed theories per se blocks scientific knowledge
about the world therefore would amount to denying the status of scientific
knowledge to all statements of modern high energy physics.

In order to avoid the latter conclusion, it seems necessary to change the
understanding of the role of assessments of limitations to scientific underdetermination in fundamental physics. Particle physics suggests a new picture
where those assessments have turned into an important part of scientific reasoning that contributes decisively to the confirmation of scientific statements.
Thereby they are constitutive of the acquisition of scientific knowledge. If
they indeed play that role in empirical confirmation, however, there is no
conceptual reason for denying the status of scientific knowledge to theoretical
claims whose characteristic empirical predictions have not been empirically
tested but which are considered very likely based on assessments of limitations
to scientific underdetermination.

A position along these lines does not deny the important difference between
theoretical reasoning and empirical testing. Neither does it deny that empirical
testing must be the ultimate goal of natural science nor that empirical confirmation substantially enhances the trustworthiness of an individual scientific theory.
It does deny, however, that there is a clear-cut distinction between viable
scientific knowledge that is always based on empirical confirmation and empirically unconfirmed speculation that is rendered tentative and unstable due to the
fact that it has to be based on reasoning open to the threat of scientific underdetermination. Rather, empirically confirmed and empirically unconfirmed
theory building should be placed on a continuum with respect to their trustworthiness. Empirically well-confirmed theories are very high up on that continuous scale under today's circumstances. It is not always true, however, that
claims based on purely theoretical reasoning are at the bottom end of the scale.
Though present-day physics does not offer good examples for a scenario of that
kind, it could happen in principle that certain statements argued for convinc-
ingly on entirely theoretical grounds may be considered more trustworthy than
others that are taken to have empirical confirmation, but in a theoretically
unclear context.

5.6 A brief excursion into palaeontology

The previous sections have shown that non-empirical theory assessment constitutes an integral part of scientific reasoning that contributes to the generation
of knowledge about the world. In the context of physics, this role has remained

under-emphasized for a long time and was hidden behind the received view of scientific progress that focusses on empirical confirmation. In this section, it shall be pointed out that other scientific fields show a decidedly different attitude.

As an example, I will examine the case of palaeontologists who try to reconstruct extinct animals from excavated fossils. Their theories are concerned with the physiology and behavior of the animals whose fossils have been found. They seek to develop theories that give full descriptions of those qualities for specific animal species.

Let us imagine the discovery of a dinosaur fossil fragment that is not attributable to any of the known dinosaur families. Let us further assume that some palaeontologist develops a theory about the physiology and living habits of the new dinosaur species based on the excavated fossil. What are the criteria based on which the scientific merits of the theory are being determined? From a physicist's perspective, the answer would be straightforward. A new theory must imply scientific predictions which can be tested. Those tests eventually are decisive for determining whether or not the theory is viable.

In the case of palaeontology the situation is not so simple, however. Most material of relevance for paleontological theory building has decayed without leaving a recognizable trace. Thus, while paleontological theories deal with observables, only a small subset of the observable objects relevant to the science are empirically accessible. A theory may be false without being refutable by any evidence accessible today or at any later point in time. It might be the case that the fossil fragment found in our example is the only part of a species of the given dinosaur family that has survived on earth, so that scientists cannot possibly get more direct information about the shape of that animal species than what is contained in it. (New contextual evidence might nevertheless change the picture, of course.) The palaeontological analysis today thus must be carried out within a framework that guarantees neither that all true scientific statements can be confirmed by empirical evidence nor that scientific statements which contradict the full body of past empirical evidence can ever be disconfirmed. Furthermore, scientists do not have clear indications as to what extent the hypotheses developed will be testable by empirical data. Still, the scientists have to come up with principles according to which they can acknowledge the scientific value of the theories they build based on the available data.

A theory that conjectures the entire physiology and behavior of a dinosaur based on minimal evidence, let us say one tooth, may be so dramatically underdetermined that most of its claims, though possibly never refutable by empirical evidence, must be expected to be false. Palaeontologists clearly must ot accept such presumably false theories and should prefer to abstain from dorsing any exhaustive theory under such circumstances. On the other hand, they would restrict the range of theory building in a counterproductive way if they allowed theory building only to the extent they knew in advance that empirical testing of the theory's claims can be provided. The method to be deployed in order to overcome this impasse is assessment of scientific underdetermination. In order to understand the scientific value of a theory that was developed based on the available evidence, the cogency of the reasoning leading to that theory has to be analyzed. In other words, it must be assessed to what extent alternative theories may exist which are compatible with the available evidence as well. In the best cases, assessment of scientific underdetermination will create a no alternatives argument. More frequently, it will lead towards a relatively small number of promising theories. Out of that group, inference to the best explanation will select one theory. That theory may, if the selection seems particularly convincing and a sufficient level of agreement within the scientific community is achieved, be taken to constitute scientific knowledge about the world.

To return to our example, an excavated dinosaur tooth may allow the conclusion that the dinosaur was a carnivore. The dinosaur evidence found so far suggests that carnivores always have a number of other characteristics, which are, on that basis, then attributed to the animal in question. Looking at patterns in the known dinosaur fossils thus provides a method for conjecturing properties of the animal which are not immediately confirmed by the evidence. An assessment of the density of evidence regarding the overall spectrum of dinosaurs then provides a basis for assessing the cogency of those extrapolations, or, in other words, the likelihood of alternative interpretations of the available fossil fragment. It may happen that scientists come to the conclusion that there probably is no scientifically reasonable interpretation of the overall findings that does not infer the stated properties. If so, scientists feel confident to call the theory that conjectures those properties knowledge about the world.

Theoretical analysis thus not just improves the factual understanding of the case but also contributes to a better grasp of the space of possibilities and therefore of the contingencies involved in modeling. The overall analysis eventually establishes a gradual distinction between unscientific "storytelling," speculative but legitimate hypotheses, and well-established theories. Describing the situation in terms of a dichotomy between empirically confirmed theories and unconfirmed hypotheses would not do justice to the gradual differences between individual cases where typically empirical data, circumstantial evidence and assessments of underdetermination merge to generate an overall judgement.

It is important to emphasize that all elements of theory assessment and theory ranking involved in the described process of reasoning got their legitimacy from past scientific successes which were based on the same methods. In other word* those methods are testable empirically in a way that closely resembles the empirical tests used in MIA. To remain faithful to our example, let us imagine that strategies similar to the ones which have led them to believe in a number of theories about the animal that owned the discovered tooth have also led them to believe in various theories on other animals based on discoveries of other fossils. If some of those other theories end up being vindicated by later findings of more complete fossils, this not only supports those theories themselves but also the methods which generated them. The increased trustworthiness of the involved methods, in turn, also raises the trust in theories which are based on our dinosaur tooth.

None of that comes as a surprise for analysts of theory assessment in the special sciences. Investigations of inference to best explanation have aimed at establishing the mechanisms behind the described kind of reasoning (Lipton, 2004; Bird, 2007). But what are the reasons for the conspicuous difference between physics and many special sciences with regard to their respective understanding of theory assessment? Why is it that assessments of underdetermination look particularly difficult and problematic in fundamental physics and are, in that context, traditionally denied the capacity of creating scientific knowledge?

Two important reasons come to mind. First, the importance of assessments of underdetermination seems to be inversely proportional to the corresponding theory's predictive capacities. Theories which produce specific and testable empirical predictions can be assessed based on those empirical tests. The question of underdetermination in that case may be neglected in considerations about the theory's status. Empirical testing trumps assessments of underdetermination. Special sciences like palaeontology are less predictive than fundamental sciences and have a more significant descriptive element. Rather than using universal natural laws to predict specific processes, they develop local schemas to describe the contingent realization of living organisms and the like. Theory assessment therefore cannot rely solely on direct empirical testing of its statements. It has to use all available means of establishing trust in a theory. Assessment of underdetermination is one strategy to that end.

Considering this point, it seems quite natural that contemporary high energy physics moves closer towards the conception of theory assessment prevalent in the special sciences. Since the testing of predictions has become more difficult to attain and theoretical contexts abound where such tests don't look realistic in the foreseeable future, the field finds itself in a situation comparable to fields like palaeontology: all other means of achieving theory assessment must be exploited. Assessments of limitations to underdetermination thus enter the picture.

Another difference between fundamental physics and the special sciences is of importance for explaining the different approach towards assessments of limitations of scientific underdetermination: the two kinds of scientific disciplines can determine underdetermination on an entirely different basis.

In fields like palaeontology, the entire reasoning is based on a stable foundation of physical and chemical knowledge. Those two disciplines provide the elementary building blocks which may be used in the palaeontological reconstruction of animal species. Palaeontologists can safely rely on the assumption that dinosaurs, just like present-day animals, consisted of the known chemical substances with the known characteristic chemical and physical properties. It just would not constitute acceptable science within the context of palaeontology to dissolve an incoherence in one of its theories by introducing altered physical or chemical laws. The existence of a conceptual foundation in the form of more fundamental scientific fields provides a framework within which the spectrum of theoretical options for theory building can be discussed. Assessments of limitations to scientific underdetermination are carried out within this framework. The existence of such a framework by no means renders the question of scientific underdetermination a simple kind of analysis. The claim of an exhaustive analysis of possible alternatives still remains difficult to argue for and constitutes one of the most intricate aspects of scientific reasoning. The most elementary foundation for that claim, however, is provided by other scientific disciplines.

Fundamental physics, to the contrary, faces the problem that no other science can offer an ontological or conceptual basis for its analysis. All elementary concepts in physics are up for change. A full understanding of the meaning of assessments of scientific underdetermination in fundamental physics thus must provide a framework for such assessments that is not rooted in any other scientific discipline. If the string physicist makes the claim that string theory seems to be the only possible unified theory of gravity and nuclear interactions, she cannot mean: the only scientific theory based on our present laws of physics. She must rather mean: the only scientific theory whatsoever. The analysis therefore must rely on general conditions of "scientificality," as they have been introduced in the beginning of this book. Unlike a framework provided by other scientific disciplines, "scientificality conditions" are meta-scientific – one may say philosophical – concepts. Unlike in the special sciences, allusions to limitations to scientific underdetermination in physics thus open up a philosophical frontline many scientists may not want to address. This in turn constitutes a significant reason why assessments of underdetermination have a dubious reputation in the field of physics. After the second half of the twentieth century had witnessed a decreasing relevance of philosophical reasoning for the working physicist, the growing importance of non-empirical theory assessmen may thus be understood in terms of a resurgent significance of philosophical reasoning in contemporary fundamental physics.

5.7 A new understanding of theory assessment

It is time to recapitulate the new understanding of theory assessment that has emerged up to this point. The fact that scientific underdetermination is limited is one of the core messages of scientific research in general and of physics in particular. This fact has been strengthened time and again by the steady sequence of successful predictions of new phenomena in science. The fact that novel predictions did not disappear but became more powerful once intuitive ontology lost its role as a guiding principle to microphysical theory building suggests that strong limitations to scientific underdetermination can be implemented based on requirements of scientificality in conjunction with consistency arguments.

Taking all this into account, it becomes strikingly clear that the canonical reconstruction of scientific theory assessment in physics is inadequately narrow. The exclusive focus on theory confirmation by empirical data that is predicted by the theory in question neglects the importance of assessments of limitations to scientific underdetermination. It is important to understand that the assessment of scientific underdetermination is by no means detached from observation. Rather, it is based, first, on evidence that is placed at a meta-level and characterizes the research process and, second, on evidence that is predicted by other physical theories. These kinds of evidence differ substantially from "empirical" evidence for a given theory, however, as they do not belong to the class of data predictable by the theory to be confirmed in the given context.

Closer analysis shows that assessments of scientific underdetermination constitute an integral part of empirical theory confirmation itself. They play a crucial role in the confirmation of unobservable objects and are essential for generating trust in the predictive power of advanced microphysical theories. This non-empirical element of theory confirmation constitutes an important extension of the canonical understanding of theory confirmation in the philosophy of science. The new, wider form of theory confirmation relies on a vague estimate of the number of possible options for alternative scientific theory building in the field. Only in cases where those options are understood to be limited, meaningful theory confirmation can take place. Where it does not, no stable concepts of microphysical objects can be developed and scientific predictions must deflate to the application of simple inductive reasoning at a phenomenal level.

It is an immediate consequence of the importance of assessments of scientific underdetermination within the process of empirical theory confirmation that the very same strategy must also influence the assessment of a theory's viability in cases where empirical confirmation for that theory is absent. We are thus led to the conclusion that non-empirical theory assessment must constitute theory confirmation on its own. This aspect of the scientific process is only of marginal importance, however, as long as we are confronted with a scientific environment where either empirical confirmation is attainable within reasonable periods of time or the basis for assessments of scientific underdetermination is weak. In the first case, it seems more reasonable to wait for empirical confirmation instead of trusting the more risky strategies of non-empirical theory assessment. In the second case, non-empirical theory assessment would be desirable but cannot provide significant theory confirmation on its own. It may be fair to say that the assessments of theories in theoretical physics from the early twentieth century until the 1980s by and large fell into one of those categories. The situation has changed substantially in recent decades, however. The increasing difficulties of contemporary theories in fundamental physics in achieving empirical theory confirmation within reasonable time frames in conjunction with the rising conceptual power of consistency arguments has led to a situation where the acknowledgement of assessments of underdetermination has become indispensable for understanding the way scientists choose, develop and trust their theories. This brings us full circle to string theory, which provides the most striking example of the power of assessments of scientific underdetermination today. In its context, the mechanisms of non-empirical theory assessment can be analyzed nicely, as we have seen in Part I of this book.

The sharp differences of opinion within the physics community about the current status of string theory and some other theories can now be explained in terms of a paradigm shift regarding the understanding of scientific theory assessment. In the physics community, those guided by the canonical paradigm of theory assessment stand against those who, led by the dynamics of their research field, endorsed a continuous increase of the role of non-empirical theory assessment. In contemporary high energy physics and cosmology, assessments of scientific underdetermination have implicitly assumed the role of a scientifically legitimate method for establishing a theory as probable scientific knowledge about the world. The traditional paradigm of theory assessment, to the contrary, is based on the strict dichotomy between scientific knowledge established by empirical confirmation on the one hand and mere speculation on the other. One important reason for the communication problems between string theorists and their critics now becomes clear. Since assessments of scientific underdetermination have always been an integral part of scientific reasonir string theorists and other physicists in high energy physics model building and cosmology can legitimately say that they are just strengthening an element of the scientific practice they had always been following. From their perspective, no substantial paradigmatic change is happening. On the other hand, critics of string physics and the – in their eyes – overly confident belief in modern cosmological theories can feel justified to reject the trust showed by those theories' exponents out of hand as unscientific because the canonical – but, as we have seen, overly simplistic – paradigm of scientific theory assessment does not allow for it. The inability of many adherents of the canonical paradigm of theory assessment to understand what physicists in string theory and other fields of contemporary fundamental physics are doing in this light appears as a late consequence of long standing deficiencies of the canonical paradigm itself. This does not imply that more specific arguments against the very optimistic outlook of some string theorists must be false. (The risks of non-empirical theory assessment were addressed in Section 3.1.) However, the foundations for principled criticism of string theory based on a comparison with the canonical paradigm now seem fairly weak. By evaluating string theoretical reasoning based on an uncompromising application of the canonical paradigm of theory assessment, that principled criticism not only denies the historicity of the principles of theory assessment, it also relies on a specific element of the canonical paradigm that has been substantially flawed all along.

PART III

Physics and truth

Part I of this book addressed aspects of theory assessment in string physics which could be understood in terms of the strengthening of tendencies that reach back far beyond the advent of string theory. Part II discussed those aspects in a wider physical framework and emphasized the high degree of continuity between string physics and earlier stages of physical progress. The third and final part of this book returns to string theory to analyze a slightly different but related topic: the status of final theory claims which arise in string physics. In that context, string theory introduces entirely new arguments and an entirely new perspective that sets the theory apart from all previous physics. However, it will turn out that the support for final theory claims relies heavily on the arguments discussed in the earlier parts of the book.

6

Final theory claims

6.1 Two final theory claims in string physics

String theorists tend to assume that a coherent formulation of string theory would constitute a final theory: no other more fundamental or more universal theory than string theory would ever be required to account for new empirical data.

The way final theory claims are discussed in string physics is in some respect reminiscent of the role played by non-empirical assessments of the theory's viability. Final theory claims clearly contribute to the fascination of string theory and arguably constitute a significant element of the overall mindset behind string physics. However, they seem to lie a little beyond the concerns of down-to-earth science, which is why their analysis mostly remains at an informal level and is more conspicuous in private communication and in publications aimed at a wider public than in scientific articles. The final theory claim was most famously discussed in Stephen Weinberg's book *Dreams of a Final Theory* (Weinberg, 1992); it is also addressed in Greene (1999) and in Kaku (1997) and has been emphasized at several instances by Hawking (see e.g. Hawking and Mlodinov, 2010).

Before analyzing the motivation for those claims, let us clarify their character by delimiting them from other finality claims which occur elsewhere in the literature. First, it is important to note that the string theorists' final theory claim is weaker than the one deployed by David Chalmers (1996) and others who deal with the perspectives for a unified description of physics and mentality. The final theory claim put forward by string physicists is confined to physics and does not address questions of mentality at all.

The perspective on final theory claims we are going to analyze also differs from the one chosen by Nicolas Rescher in recent work. Rescher discusses a final theory concept that is based on the idea of an ultimate explanation (see Rescher, 2000). According to that understanding, a final theory must fulfil two conditions. First, it must be universal, which means that it covers all facts which arise in this world. And second, it must provide the "deepest possible explanation" for each of those facts. Rescher then goes on to discuss consistency problems which arise with regard to this conception of finality.

Rescher's understanding is clearly related to the kind of final theory claims discussed in string physics. Nevertheless, I will choose a slightly different perspective that seems easier to define in a string theoretical context and also avoids some of the problems faced by Rescher. That notion of a final theory avoids Rescher's reference to a "deepest possible explanation" and instead is based on two core concepts: the condition of scientificality and the range of a theory's empirical implications. A final theory claim in the given sense asserts that the theory alleged to be final does not have any possible rivals that provide more concise or universal predictions of empirical data and fulfil the scientificality conditions.

Rescher's notion of a "deepest possible explanation" is avoided for two reasons. First, it seems conceptually very difficult to grasp. And second, it directly leads to the consistency problems identified by Rescher himself with respect to his notion of final theory claims. The notion of scientificality conditions seems less problematic in these respects. Scientificality conditions were introduced in Part I in order to provide a framework for statements of scientific underdetermination. There, it was asserted that only a limited number of theories which satisfy the scientificality conditions and are distinguishable by a certain set of experiments could be built. Now, the same conditions are used as the basis for a final theory claim. On that basis, the analysis of final theory claims can remain purely empirical. They are comparably easy to grasp and do not create the consistency issues raised by Rescher.

In the given definition of a final theory, scientificality conditions are used for avoiding problems related to the existence of correct pure data-models. In case of a non-deterministic final theory – which is a plausible scenario in a quantum physical context – it is safe to assume that there exists one data-model that correctly reproduces every single piece of data ever to be collected in this world. (Of course our chances of finding this model are essentially zero.) Such an "omniscient" data-model gives precise predictions for all experimental outcomes and thus is more precise than any irreducibly stochastic theory. In order to be able to make a final theory claim for a stochastic theory one thus has to exclude such omniscient data models. In the chosen approach, they are indeed excluded because they do not satisfy an important scientificality condition: they involve lots of ad-hoc assumptions for modeling individual events.

The suggested definition of finality goes beyond the claim that a scientific theory is compatible with all possible data but remains a little weaker than definitions based on deepest or ultimate explanation. A theory that is capable of reproducing all possible data but leaves undecided some questions regarding experimental outcomes which are correctly predicted by another theory is no final theory in the given sense. On the other hand, the existence of a final theory in the given sense does not exclude the existence of other, empirically equivalent theories which are based on different mathematical principles and thereby offer a better overall understanding of the interconnections between the theories' various conceptual aspects. Rescher's notion that a final theory provides the deepest possible explanation would imply that alternatives of the latter kind are absent as well. In the case of string theory, however, the theory's theoretically incomplete character in conjunction with the important role of duality relations renders questions of the latter kind particularly difficult to grasp. It may be futile to distinguish between distinct theories on the one hand and mere distinct formulations on the other in the given context. For that reason, it seems advisable to restrict the discussion on finality to statements on empirically distinguishable alternative theories.

String theorists have two main reasons for believing that their theory, if valid at all, might be final. First, string theory is the first physical theory that seriously claims to provide a fully unified description of all known fundamental physical phenomena. If string theory is a viable theory, all fundamental physical phenomena can on its basis be described by one integrated set of physical principles. To understand the status of this claim, it is helpful to put it into a historical perspective. After Maxwell had developed the classical form of electrodynamics, some observers of physics in the 1880s believed that Newtonian physics (i.e. classical mechanics plus Newtonian gravity) and electrodynamics were at the brink of providing a full description of all known physical phenomena. (It was hoped that the few remaining anomalies could be resolved in the near future.) The conjunction of Newtonian physics and electrodynamics was no fully unified description of all phenomena. After all, the two main theories relied on quite different and seemingly unrelated physical principles. However, it was by no means clear at the time that full unification constituted a relevant goal of physics. If two disjoint theories could cover the entire spectrum of observable phenomena, that could be taken to be a fully satisfactory final description of nature.

Today, we know that the hopes for an imminent closure of fundamental physics at the end of the nineteenth century were unfounded. The changes with regard to our fundamental understanding of space, time and objecthood which went along with the development of relativity and quantum physics gradually eroded the previously stable intuitive basis for physical theory building. That intuitive basis, however, had been the place where independent but compatible theories like electrodynamics and mechanics could meet. It was the picture of solid and well-located objects moving in three-dimensional space that provided the link between Newtonian mechanics and Maxwell's electrodynamics. By being massive, these objects felt the gravitational force. If charged, they could at the same time serve as test particles for the predictions of electrodynamics. The new physical perspective that lacked such a universal intuitive basis turned the coherent description of a spectrum of different phenomena into a more intricate enterprise. Full coherence now was difficult to achieve since even the most fundamental properties of objecthood, space and time depended on the individual theory. Theories which did not share one common concept of space and time or localization became very difficult to combine in a coherent way. Under the new circumstances, the requirement of full coherence thus was closely related to full unification. The latter therefore became a natural goal of physical theory building.

String theory unifies nuclear interactions and gravity and therefore, if valid, would cover all physical interactions known today. If full universality is a goal of physical theory building, string theory thus could be the theory that terminates the sequence of theories that were coming closer to its fulfilment.

A far more powerful final theory claim is related to an interesting implication of string dualities. The argument has e.g. been presented in a cogent way in Witten (1996). The string world shows a remarkable tendency to link seemingly different string scenarios by so-called duality relations. As discussed in Chapter 1, two dual theories are exactly equivalent concerning their observational signatures though they are constructed quite differently and may involve different types of elementary objects and different topological scenarios. The kind of duality relation relevant to our present context is T-duality. Let us briefly repeat the basic characteristics of that concept. String theory suggests the existence of compactified dimensions. Closed strings can be wrapped around compactified dimensions like a closed rubber band around a cylinder. They can also move along compactified dimensions. Due to the basic principles of quantum mechanics, momenta along closed dimensions can only assume certain discrete quantized eigenvalues. Thus, two basic discrete numbers exist which characterize the state of a closed string in a compactified dimension: the number of times the string is wrapped around this dimension, and the eigenvalue of its momentum state in that very same dimension. Now, T-duality asserts that a model where a string with characteristic length l is wrapped n times around a dimension with radius R and has momentum eigenvalue m is dual to a model where a string is wrapped m times around a dimension with radius l^2/R and has momentum eigenvalue n. The two descriptions give identical physics.

This fact can be generalized and eventually implies that all tests which seem to be directed at distances smaller than the string length can be understood as tests of correspondingly larger distances. Duality thus translates all information below the string length into information above the string length, rendering the former fully redundant. The testing of ever smaller distance scales in a sense "bounces back" at the string scale and ends up measuring what is most adequately described as larger distance scales once again. Since no new degrees of freedom open up below the string scale, it makes sense to speak of a minimal length in physics. The absolute limit set on attaining new physical information below a certain scale formally puts an end to the continuous physical search for new phenomena at ever smaller distance scales. String theory implies that one cannot go beyond it by looking closer at nature. The internal structure of the theory thus contains a final theory claim.

The string-theoretical posit of a minimum length must be clearly distinguished from weaker finality assertions which appear in earlier physical theories without justifying a final theory claim. For example, the posit of elementary particles in microphysics may have an air of finality but does not exclude independent new phenomena which require additional theory building. It therefore merely establishes a temporary concept of fundamental elements without implying any final theory claim that would assert that these elements must remain fundamental forever. A limit to the physically achievable values of a physical parameter is set by the absolute speed limit in special relativity. However, this speed limit does not translate into a finite limit to the kinetic energy per particle. Therefore, since any experiment must be carried out at a finite energy scale, special relativity always leaves room for new physics beyond that scale. String theory is the first theory where a universal limit to possible new physics is implied by the theory's structure and where physics up to that limit could in principle be tested by experiment.

Critics of string theory often consider final theory claims unscientific. The general sentiment behind critical remarks published in some books and articles (see e.g. Dyson, 2008), is arguably shared by many physicists and philosophers of science. We will not carry out an exhaustive analysis of all arguments which have been developed in this context. In keeping with the strategy pursued throughout this book, we will rather focus on the aspect of epistemic access. At that level, we find a rather straightforward line of reasoning against the viability of final theory claims. In a nutshell, the argument works the following way: even if one granted the theoretical possibility that a final theory existed, it seems that no sound epistemic basis can justify attributing that status to any specific theory. A fundamental obstacle seems to block the epistemic access to finality: while final theory claims address the question of new theories superseding the present one, all arguments which can support them are necessarily based on the currently available data and our present web of theories. Accepting that framework as absolute, however, means begging the question of finality. Any final theory claim in this light seems to be based on an unfounded denial of the historicity of scientific reasoning.

Let us spell out this problem more specifically with respect to the two final theory claims described above. The minimal length scale implied by T-duality allegedly supports string theory's status as a final theory. But the minimal length scale is itself derived within the framework of string theory. If string theory were a viable theory but constituted only an effective description of an even more fundamental theory, that more fundamental theory would confine string theory's viability to a certain regime (presumably roughly up to its characteristic scale, the string scale). String theory's assertion of a minimal length scale, by talking about what lies beyond the string scale, would then address a regime where string theory is no longer valid. It could still be the case that the more fundamental theory, though being consistent with all conceptual implications of string theory up to the string scale, does not contain a minimal length scale. The final theory claim thus seems to fail.

If the final theory claim cannot be supported by conceptual arguments, the statement on string theory's full universality loses much of its appeal as well. Obviously, the latter statement can only be formulated with respect to the set of phenomena known at the time. Without support from a conceptual final theory claim like the claim of a minimal distance scale it thus cannot rule out that new phenomena which reach beyond the allegedly universal theory might be discovered in the future. The argument from universality then seems to deflate to the mere statement that the present theory manages to describe all phenomena presently known.

6.2 Local and global limitations to scientific underdetermination

Looking a little closer at the stated argument against final theory claims, one realizes that it is embedded within a specific frame of thought: it is based on the canonical paradigm of theory assessment according to which a theory's status can only be assessed to the extent it has been tested by empirical data. It is on that basis that the argument is convincing: since the available data will never exhaust the data that could be collected in principle, the canonical paradigm implies that judgements regarding the status of scientific theories can never address the question of finality. The earlier parts of this book have demonstrated, however, that the canonical paradigm of theory assessment does not square well with the actual research process in contemporary high energy physics. In contradiction to the canonical paradigm, assessments of limitations to scientific underdetermination do reach beyond the horizon set by the currently available empirical data. Assessments of scientific underdetermination thus demonstrate that the classical epistemic argument against final theory claims is not conclusive. Moreover, it seems natural to suspect that assessments of limitations to scientific underdetermination themselves may play a crucial role in the strengthening of final theory claims. The following will analyze to what extent these promises can be fulfilled.

Is it possible to address the final theory case along the lines of reasoning discussed in Parts I and II of this book? At any rate, an extrapolation of that kind is not entirely straightforward. As discussed in Section 3.4, the case of final theory claims differs substantially from the case of non-empirical theory assessment presented in earlier parts of this book. Part I established the existence of rational arguments for what I have called "local" limitations to scientific underdetermination. A scarcity of alternatives was argued for with respect to bundles of theories which gave distinguishable predictions regarding some limited set of future experiments. A final theory claim, to the contrary, must be based on an assessment of "global" limitations to underdetermination, that is, limitations to the number of theories which are empirically distinguishable based on all possible observations. If we want to use the arguments presented in Part I for supporting final theory claims, we therefore have to establish that, under appropriate circumstances, assessments of local limitations to scientific underdetermination can support the case for global limitations as well.

Let me first pinpoint a subtle aspect of local scientific underdetermination in order to make the framework of the analysis a little more explicit. Local scientific underdetermination is instantiated by the set of alternative theories which are compatible with the available data and are not empirically equivalent with regard to the next stages of empirical testing. These alternative theories are characterized by spectra of possible empirical outcomes they allow for. The question as to whether or not an individual theory contributes to local underdetermination remains untouched by the question whether the theory is actually realized in nature and, if so, with which parameter values. A theory that, at a far later stage, will actually turn out to be realized with parameter values that render it indistinguishable from the presently endorsed theory at the next stages of empirical testing therefore can still contribute to local underdetermination today. It does so if, given the presently available data, it *allows* for parameter values which *would* make it distinguishable at the next stage of empirical testing. Finding theories which contribute *only* to global scientific underdetermination in this light is a little more difficult than one might think at first glance. It requires finding theories that are not empirically equivalent with the other possible theories but whose empirical deviations from some other possible theories could be known already at this point to be too small to be detected at the next stages of empirical testing.

Under "normal" scientific circumstances, one can indeed find theories which satisfy these conditions and therefore exemplify purely global scientific underdetermination. Therefore, *local limitations* to underdetermination can usually be established without implying any *global limitations*. In special sciences, this is possible due to what one could call the vertical hierarchy among scientific theories. A clear distinction can be drawn between the characteristic length scale of some phenomenon to be explained by a theory in a special science and smaller length scales that characterize more fundamental scientific theories on which the theory in question is implicitly based. Thus, even if only few mechanisms can explain the available data and provide distinguishable predictions for future experiments at one level of description (which means that local underdetermination is limited at this level), many different realizations of those mechanisms may be possible at the micro-levels addressed by more fundamental theories. The spectrum of possible lower-level mechanisms is often irrelevant to the scientific discourse at the higher level. The selection of the mechanism at the lower level influences neither the empirical predictions nor the conceptual analysis at the higher level. Therefore, this spectrum of lower level mechanisms does not contribute to local scientific underdetermination but contributes to global underdetermination. Without knowing more about that spectrum, no conclusions regarding global limitations to scientific underdetermination can be drawn.

To give an example for a research situation of that kind, let us once more look at a case from palaeontology. Evolutionary biologists have suggested a number of evolutionary mechanisms that may have led to the development of wings. They may argue for the viability of an individual theory about the evolutionary development of wings in a specified context by pointing out that the theory in question seems to be the only theory that can account for all known empirical evidence relevant in the given context. In doing so, they make a claim of local limitations to scientific underdetermination. Now, any modern palaeontological or biological theory is implicitly based on the principles of modern physics. Any theory on the evolution of wings thus subscribes to the understanding that animal wings consist of atoms which, in turn, are built of more fundamental elementary particles. By claiming local limitations to scientific underdetermination, evolutionary biologists do not address these fundamental levels of fundamental physics. They do not claim that only one set of fundamental physical theories can underlie the conjectured macroscopic evolutionary mechanism. The question as to which microphysical theories provide a viable description of atoms and elementary particles is not a subject of the initial biological investigation and has to be established through experiments at entirely different energy scales carried out in microphysics. Local limitations to scientific underdetermination thus do not translate into global ones in the given case.

In fundamental physics, the distinction between local and global underdetermination often works in a similar way as in the special sciences. The role played by the hierarchy of theories in the special sciences is now taken over by the limited range of viability of individual theories within a more universal general conception of physical phenomena. We may know about a theory B which constitutes an alternative to a theory A but which is characterized by a certain parameter that has been measured already within a physical context unrelated to theory A. Let us now assume that the measured parameter value lies far beyond the range of the next generation of experiments E which have been planned in order to test theory A. In that case, we know that theory B, while relevant for our understanding of global underdetermination, remains irrelevant for local underdetermination with respect to experiments E: theory B does not constitute an alternative theory that, according to our best knowledge, can be distinguishable from A by experiments E.

To give an example of this way of reasoning, let us look at non-relativistic quantum mechanics. Quantum field theory constitutes a theoretical alternative to non-relativistic quantum mechanics that becomes empirically distinguishable from the latter once processes close to the velocity of light enter the picture. Physicists at the time when quantum physics was developed knew the speed of light from astronomical measurements which had no connection to quantum physics. That knowledge fixed the scale where non-relativistic quantum mechanics and quantum field theory became distinguishable and, on that basis, created a regime where quantum field theory was known not to constitute an empirically distinguishable alternative to quantum mechanics. Therefore, quantum field theory left local limitations to underdetermination within that regime untouched and only contributed to global underdetermination.

Having established the nature of the distinction between local and global underdetermination in fundamental physics, I now come to the core of the present analysis. Contextual arguments like the three arguments for limitations to scientific underdetermination presented in Part I of this book on their own cannot touch the question of global scientific underdetermination for a general structural reason. The processes of theory development and empirical testing on which they rely as their evidential basis are themselves of a local character. It would be unconvincing to infer a theory's unlimited viability from predictive success within a limited regime or from a lack of alternatives found within a limited time frame. Therefore, the three arguments themselves cannot be promising candidates for establishing final theory claims. On the other hand, the argument against final theory claims presented in Section 6.1 has established that the canonical final theory arguments as deployed in the context of string physics must remain unsound without further conceptual support. The idea now is to join the two spheres in order to build one overall line of reasoning that amounts to a significant final theory claim. It will turn out that the two final theory arguments presented above can be understood as arguments which block the distinction between local and global underdetermination. On that basis, the arguments for limitations to local scientific underdetermination can establish the final theory claim.

Before starting the analysis, it is important to specify the limit between local and global underdetermination. It makes most sense to put that limit at the characteristic scale of the theory to be analyzed, where its core predictions can be tested. In the case of string theory, that is the string scale. It would obviously make no sense to discuss local underdetermination with respect to the next generation of collider experiments which most likely are incapable of testing string theory at all.

Now, consider the case of a fundamental theory that is fully universal in the sense that the theory covers all available conceptual information on parameter values which characterize known phenomena. In other words, physicists are not aware of any physical parameter that neither occurs in the universal theory itself nor can be reduced to parameters of the universal theory. In a scenario of that kind, any parameter that has already been measured and controls the extent to which the universal theory and a possible alternative theory are empirically distinguishable must play a role within the universal theory. The mechanism presented in the previous paragraphs as the most common basis for separating global from local scientific underdetermination, however, crucially relied on parameters which played no role in the theory under consideration (e.g. the velocity of light in the case of non-relativistic quantum mechanics). That strategy therefore breaks down in the given context, which significantly reduces the options for alternative theories that do not affect local limitations to scientific underdetermination.

Alternatives which only count at a global level thus can only arise if a constellation *within* the known universal theory constrains alternative theories to regimes which remain inaccessible to those experiments that test the present theory's core predictions. An example for such a constellation would be the relation between the standard model and grand unified theories. In order to be compatible with present empirical data, grand unification must occur at the energy scale where the known gauge couplings (which have been measured close to the electroweak scale and can be calculated for higher energy scales based on renormalization group techniques) all assume the same value. This scale, called the GUT scale, is known to lie far beyond the grasp of collider experiments, however. (We want to disregard the possibility of large extra dimensions for the moment.) On that basis, we can know that grand unified theories cannot be distinguished from the standard model in collider experiments which test energy scales far below the GUT scale. Grand unified theories thus constitute alternatives which affect global underdetermination but most probably are of no concern to local underdetermination immediately above the electroweak scale.

Constellations of the kind described in the last paragraph, however, are restricted by the second final theory argument, the argument for a minimal length scale based on duality. Once such a minimal scale is established within a universal theory at that theory's characteristic scale, the statement that this theory remains viable beyond its characteristic scale but stops being viable far beyond that scale does not make sense within the theory's framework. Neither the theory itself nor any specific choices of its parameter values thus can enforce a detachment of the scale where new physics becomes relevant from its own characteristic scale. If new physics existed, nothing in our present theory could prohibit it from becoming observable immediately beyond the theory's characteristic scale. In other words, a theory that implies a minimal distance scale cannot enforce the splitting between local and global scientific underdetermination.[1]

Both final theory claims presented in the context of string theory in this light may be understood in a new way. On their own, the two arguments cannot establish that theory succession which goes beyond string theory is unlikely. They do establish, however, that the question of global underdetermination conflates with the question of local underdetermination. Therefore, arguments which suggest limitations to local scientific underdetermination must be acknowledged as arguments against unlimited global scientific underdetermination as well. The final theory arguments play the role of mediators which connect the local and the global level of analysis and thereby raise the significance of local arguments. Local limitations to scientific underdetermination in contemporary high energy physics thus can be shown to translate into global limitations and final theory claims can attain a certain degree of trust.

---

[1] It should be emphasized once more that this does not imply that the alternative theory must in fact have observable implications immediately beyond the string scale. It just means that, if it has any free parameters, it must allow for parameter values which make it empirically distinguishable immediately beyond the string scale.

The presented line of reasoning relies on two hidden presumptions which must be addressed in order to determine the force of the resulting final theory claim. First, the described reduction of global to local underdetermination is only implemented with respect to distance and energy scales, which, as discussed above, define the range of experimental testing in high energy physics. This does not necessarily imply a full reduction of global to local underdetermination since other parameters could control a theory's empirical testability. For example, the detection of small deviations from a predicted statistical distribution may require a very high number of observed events. Now the parameters which control empirical significance in those cases may be and sometimes are theoretically related to the theory's characteristic energy scale. (A good example for such dependence is proton decay, where the phenomenon's observability depends on the number of protons observed but the theoretical framework relates proton decay to the existence of new particles with very high masses.) If no such theory-based correlation with energy scales exists, however, new physics might arise without having any characteristic energy scale. In that case, the new parameter could give reason for considering global underdetermination that does not imply local underdetermination. To give an example, one could think of a theory that deviates from the statistical distributions predicted by quantum physics in a posited re-contracting phase of the universe. That theory obviously offers an alternative to quantum mechanics and thereby implies global underdetermination. However, since we know that we live at a time when the universe is expanding, we also know that the theory in question does not differ empirically from quantum mechanics with respect to any experiment or observation feasible in the near future. Therefore, the theory does not imply local underdetermination. A full assessment of the prospects of a final theory claim would have to address the chances for such scenarios as well. In this light, it is not justified to claim that the described arguments for finality conclusively rule out any perspective for strictly global underdetermination. What the arguments do is reduce the options for such scenarios, which raises the prospects for a final theory claim.

Second, it was assumed in the previous analysis that any possible successor to our present theory has continuous free parameters which control the empirical impact of that new theory. While this may have seemed a rather innocent assumption throughout most of the history of physics, it clearly has become questionable in a string theoretical context. After all, string theory itself does not contain any free parameters. This does not per se exclude the possibility that a successor theory may once again return to having free parameters. Therefore, it is still of interest to address the question of scientific underdetermination with respect to theories with free parameters. However, the emerging understanding that theories without free parameters can constitute realistic scenarios in physics opens up a new dimension of analysis. Such theories enforce discrete points in parameter space. They introduce a new scenario for alternative theories which contribute to global but not to local scientific underdetermination. This scenario is instantiated when all discrete solutions of a theory without free parameters lie far beyond what is empirically distinguishable today from the currently endorsed theory. The theory then obviously contributes to global scientific underdetermination. Since it cannot be tested at the next generation of experiments, however, it does not contribute to local scientific underdetermination. As such a scenario implements the split between local and global underdetermination based on the new rather than the currently endorsed theory, its significance is not threatened by minimal length scales within the context of the current theory.

To assess whether or not that scenario seriously undermines final theory claims, we have to view the situation from a different angle. Eventually, we will be able to extract an altogether new final theory claim on that basis.

6.3 Structural uniqueness and its impact on the question of finality

Usually, physical theories have free parameters which can be tuned in order to fit the quantitative specifics of the empirical evidence. Special relativity does not specify the velocity of light, neither Newtonian mechanics nor general relativity specify the size of the gravitational constant, and neither Maxwell's electromagnetism nor quantum electrodynamics fix the size of the elementary charge and the fine structure constant, respectively. The standard model of particle physics, the current joint description of all nuclear interactions, involves more than 20 free parameters such as mass terms, coupling constants and mixing angles which have to be determined experimentally. String theory differs from all these examples. It is the first physical theory that does not contain or allow any fundamental free parameters (see e.g. Polchinski, 1998, Chapter 1).[2] According to string theory, all

---

[2] For the sake of precision, one should elaborate a little further on this point. Physical quantities are characterized by their basic physical dimensions like length, time or mass. Only dimensionless parameters can be varied in a physically meaningful way in order to fit empirical data. Dimensionful parameter values depend on the choice of the basic units and therefore do not carry absolute physical meaning. Some of the fundamental constants mentioned above are dimensionful, however. In those cases, the embedding of the respective theories within our observed world requires a comparison of their fundamental constants with the scales of the everyday world and dimensionless parameters arise on that basis. A theory like string theory that gives a joint description of all fundamental phenomena does not require an embedding of its characteristic scales into some "rest of the world" that is not covered by the theory. In this case, the question of free parameters is reduced to existence of dimensionless free parameters within the theory itself.

quantitative characteristics of the world stem from purely structural character-istics of the string and are a result of string theory's complex dynamics.

A similar statement can be made concerning the spectrum of possible models of the theory, which are defined by various discrete characteristics such as symmetry structure, number of particle generations, number of spacetime dimen-sions, etc.[3] Most traditional theories allow a considerable amount of structural choice within their fundamental theoretical framework. Gauge field theory, to give an example, is highly flexible in accommodating all those microscopic phenomena which happen to show up in an experiment. Theoretical arguments do not predetermine the gauge symmetry structure and the number of particle generations and therefore allow a nearly unlimited number of models with differ-ent interaction structures and particle contents. The standard model constitutes one of those possible models, the selection of which is based on the collected empirical data. Supersymmetry or GUTs introduce a constraint on the gauge structure beyond the standard model but themselves allow for a wide range of models. It is not possible at the present stage to select one specific model because the required empirical data is not yet available. Unlike all those examples, string theory seems to allow only one model under very basic conditions, such as the existence of fermions and more than one spatial dimension. Today, it is generally assumed that there is exactly one way to build a superstring theory at the fundamental level.

I will refer to the fact that string theory knows neither free parameters, nor a variety of models, by using the term "structural uniqueness." The structural uniqueness of the fundamental theory has to be clearly distinguished from the question of the string theory ground state. As described in Section 1.1, string theory is assumed to contain "compactified" dimensions, which run back into themselves after a minimal distance like a tiny cylinder surface or torus. The compactification of those extra dimensions is a matter of the theory's dynamics. A very large number of geometries of the six-dimensional compactified space (so-called Calabi–Yau spaces) can be mathematically constructed. In recent years, substantial progress has been made towards constructing stable ground

---

[3] Here, the term "model" is used in the way it appears in particle physics in notions like "standard model" or "model building." It refers to the specific theoretical constructions which are possible within some theoretical framework, i.e. a fundamental set of physical postulates. The various models of a theory are not ontologically equivalent and have different empirical implications. The distinction between "theory" and "model" emphasizes the important difference between the fundamental conceptual framework of a theory and the specific choices made when selecting the theory's precise form. As noted in Section 5.1, string theorists speak of types of string theories, which, however, are all connected by duality relations and therefore do not constitute empirically distinguishable models.

states based on models which involve D-branes and string-fluxes[4] in Calabi–
Yau spaces.[5] Relying on these techniques, it is estimated that the number of
possible stable ground states of string theory which results from the large
number of Calabi–Yau spaces and the many possible combinations of string-
fluxes and D-branes may be as high as 10^{500}. If the physical equations allow for
many stable ground states, the choice of the actual ground state would be a
matter of the statistical quantum dynamics of the early string-universe and could
not be determined theoretically.

It has always been a natural goal of string theoretical research to establish the
theory's predictive power at low energies. A huge number of possible ground
states like 10^{500} would of course substantially reduce the predictive power of
string theory. The dynamics of string theory is too little understood today to
allow any final judgement on the number of ground states which are actually
physically possible. It cannot be excluded that some kind of so far unknown
vacuum selection mechanism reduces the number of physically viable ground
states to a substantially smaller number than the one stated above. While some
string physicists consider the emergence of such a vacuum selection mechanism
a serious possibility, others have fully endorsed the reduction of predictive
power in string theory and argue that it is exactly the huge number of possible
ground states that allows for an explanation of the fine-tuning of our universe
based on an anthropic principle (Douglas, 2003; Susskind, 2003).

Irrespective of the actual number of ground states in string theory, however,
the string theoretical property of structural uniqueness changes the perspective
on degrees of freedom in physics. While the classical perspective has been
based on the notion that theories can always be fitted to nature by tuning
continuous parameters, string theory introduces the new understanding that a
scientific theory may allow just a finite number of discrete solutions. This
implies that a theory's predictive power can in principle be much higher than
what one would expect from a conventional scientific theory. Even though, as
has been pointed out, the actual predictive power of string theory is unknown at
this point, the fact that the conceptual framework of string theory turns substantially higher predictive power into a scientific possibility changes the
natural attitude towards scientific underdetermination. While it is per se natural
to assume underdetermination in a scientific context where the existence of free
parameters is taken for granted, it will turn out to be far less natural to do so in a

---

[4] String fluxes correspond to specific oscillation modes of the string which do not correspond to any
of the known types of point-like particles and had mostly been neglected in earlier string
theoretical analysis.
[5] These constructions are not exact but are based on approximations which are considered
acceptable.

scientific context where free parameters are absent and the level of predictive power depends on other more intricate specifics of the theory. The following discussion thus will not rely on the actual predictive power of string theory but will investigate the implications of higher predictive power in principle.

I will call a structurally unique theory "highly predictive" if the number of its physically possible ground states is significantly lower than the number of distinct values which could be distinguished by a precision measurement on a continuous free parameter within reasonable boundaries. Before analyzing the highly predictive case, it is instructive to take a short look at theories without free parameters that are not highly predictive. In particular, it is interesting to consider the role played by such theories in the previous analysis of the distinction between local and global scientific underdetermination. Non-highly predictive theories allow for so many discrete solutions that one may expect some of these solutions to be empirically distinguishable from our present theory at the next stage of empirical testing. With regard to the possibility that such theories contribute to global but not to local scientific underdetermination, these theories thus seem comparable to theories with free parameters. The arguments supporting final theory claims in as far as alternatives with free parameters are concerned seem to work with respect to non-highly predictive theories without free parameters as well. The case that creates a substantially new scenario is the highly predictive case, which shall be analyzed in detail in the following.

In order to carry out that analysis, one has to find a more general perspective on the question of scientific underdetermination. Large parts of this book have been spent on analyzing specific arguments which suggested that scientific underdetermination was more limited than one was inclined to believe. But let us return to a more basic question. Why do we believe scientific underdetermination to be prevalent in the first place? A brief answer to this question has already been given in Chapter 2. The principle of scientific underdetermination acquires its plausibility based on a specific understanding of the scientific process. According to this understanding, the scientist builds theoretical structures, which reflect the regularities observed in nature up to some precision, and tunes the free parameters contained in those structures to fit the quantitative details of observation. The successful construction of a suitable theory for a significant and repeatedly observable regularity that characterizes the world is assumed to be just a matter of the scientist's creativity and diligence. If it is always possible to find one suitable scientific theory, however, it seems natural to assume that there can be others as well. There may always exist different choices of theoretical structure that have coinciding empirical implications up to some precision in the observed regime if their respective free parameters are fixed accordingly. The principle of underdetermination follows from this.

One notices that this picture crucially relies on the presence of free parameters which can be tuned to fit the data. It is the flexibility of the scientific theories we are used to working with which provides the basis for making the standard understanding of scientific theory building plausible.

If one considers only the class of highly predictive structurally unique theories, one encounters an entirely different situation. Compared to the general case of all possible scientific theories, the chances for being able to describe a specific empirical data set with a highly predictive structurally unique theory are strongly reduced for two reasons. First, a highly predictive theory that does not allow any freedom of choosing parameter values or modifying qualitative characteristics in order to fit the empirical data is compatible with far fewer sets of empirical data than a conventional theory. Second, the difficulties to come up with structurally unique theories suggest that there are far fewer structurally unique theories than conventional ones. In fact, the only structurally unique theory known in science today is string theory.

Considering both the generality of the principles which define a theory such as string theory and the vast range of possible phenomenological regularities and parameter values, it is most natural to assume that highly predictive structurally unique theories, if found at all, can only be found for a small subset of points within the huge space of all possible regularities. Since there is no reason for expecting that different structurally unique theories, each of them based on a different set of fundamental physical principles, have a tendency to give similar empirical predictions, the notion that the entire set of data we have collected in our experiments can be accounted for by several highly predictive structurally unique theories which are not empirically equivalent must then be considered highly improbable. The principle of scientific underdetermination, which looks convincing if applied to the set of all scientific theories, therefore lacks plausibility if applied to the set of highly predictive structurally unique theories. If we assume that structurally unique theories can only be expected to be superseded by theories which are once again structurally unique, it therefore seems most plausible to expect that a highly predictive structurally unique theory that has found empirical confirmation will not be superseded at all.

One might still try to defend the principle of theory succession by retreating to the claim that highly predictive structurally unique theories might be superseded by theories which do have free parameters. Such alternatives with free parameters to a theory without free parameters could be of two different kinds. First, an alternative theory could be structurally unique with respect to all parameters and theoretical features relevant for the description of the phenomena which had been accounted for by the predecessor theory while introducing a free parameter that is relevant only for phenomena unknown to the predecessor theory. Since such a free parameter would not enhance its theory's chances of being coherent with the empirical data that was accounted for by the predecessor theory, however, the occurrence of an empirically viable theory of that kind cannot be considered more likely than the occurrence of an alternative highly predictive structurally unique theory. This scenario therefore leads back to the assessment of the likelihood of scientific underdetermination among structurally unique theories.

Second, alternative theories could lack structural uniqueness or high predictiveness with respect to phenomena correctly accounted for by a highly predictive structurally unique theory. The probability of the empirical success of alternative theories of that kind is not constrained by the considerations of the previous paragraphs. However, a different kind of argument can be raised against the assumption that such alternatives could be scientifically successful. If a structurally unique and highly predictive theory were able to reproduce the available empirical data at some stage, we would expect that any successor theory were able to explain the agreement between the predictions of the old theory and the empirical data. A successor theory that is not structurally unique or not highly predictive in the same regime could not provide this explanation, however. Why should a highly predictive theory be able to reproduce the empirical data without any tuning of parameters at some stage if it eventually has to give way to a theory that either requires the tuning of parameters or lacks comparable predictive force for other reasons? The fact that a highly predictive structurally unique theory had been successful at all at some stage would turn into a miracle.[6]

In this light, the question of the existence of alternative theories that are not structurally unique, though still legitimate, loses its crucial importance for the understanding of theory succession. Once we have at some stage found a highly predictive structurally unique theory that is viable at its own characteristic scale, it may be expected that no theory succession leads back to theories which do not share the property of highly predictive structural uniqueness.

Summing up the previous arguments, a highly predictive structurally unique theory that fits the empirical data at some stage can be expected to be replaced neither by another structurally unique theory nor by a theory that does not belong to that class. This suggests the termination of the progressing sequence of

---

[6] The basic argument applied here is not specific to theories without free parameters. If some theory explains the quantitative relation between two of its parameters, any viable successor theory must be expected to be able to explain that relation as well. Otherwise, the principle that a theory should be able to explain its predecessor's success would be violated. In the case of a structurally unique theory, all quantitative relations between its parameters are fixed. This means that any successor theory which has a free parameter that is relevant for the structuring of phenomena described by the predecessor theory necessarily runs into the stated problem.

scientific theories. It must be expected that a highly predictive structurally unique theory that fits the present experimental data should describe all future experiments correctly as well. Thus, one must feel compelled to call any empirically successful structurally unique theory a serious candidate for a final theory.

Note that this conclusion goes decidedly further than the final theory claims discussed before. While those merely reduced the question of global scientific underdetermination to the local question, where other arguments had to take over, the present argument establishes a fully fledged final theory claim on its own. It therefore remains conceptually independent from the validity of the three arguments for limitations to local underdetermination discussed in Part I of this book. However, the application of the argument to string theory relies, first, on the condition that string theory is highly predictive and, second, on the assumption that string theory is a viable theory at some level. The legitimacy of the latter assumption, at this point, can only be inferred based on the three arguments for limitations to local scientific underdetermination after all.

But even with respect to the nature of empirical evidence necessary for establishing a theory's viability, the conclusion reached above has an interesting implication. If a highly predictive structurally unique theory were able to reproduce the empirical data at any energy scale, this could be taken as a strong confirmation of that theory's characteristic predictions, even if those predictions had not been tested themselves and the available data could also be reproduced by other theories. In the case of conventional theories, the inference from a theory's empirical adequacy at one scale to its viability at another scale was prevented by the scientific underdetermination principle. But in the context of highly predictive structural uniqueness, that principle loses its pivotal role and the inference becomes viable as a statement about the most reasonable expectation. String theory might someday show the significance of this consideration: if it turned out that string theory delivered specific low energy predictions which fit all presently known phenomenological data in particle physics within the given experimental limits of accuracy (just like the particle physics standard model does today), this would obviously not constitute experimental confirmation of strings at their own characteristic scale. Nevertheless, even the staunchest supporter of the predominance of experimental confirmation would be obliged to attribute a very high probability to the validity of string theory's high energy predictions.

6.4 The formal structure of final theory arguments

Unlike the arguments for local limitations to scientific underdetermination discussed in Part I, the final theory arguments analyzed above do not exemplify inference to the best explanation (IBE). IBE supports a certain statement based on the claim that it constitutes the best explanation of an observed phenomenon or derived property. But none of the presented final theory arguments, neither the inference from universality nor the argument from a minimal length scale or the claim of the improbability of finding a highly predictive structurally unique alternative theory, are based on providing explanations for any features of string theory or its evolution.

In order to understand why they do not, let us, for comparison, look once more at the explanatory structure of the arguments for local limitations to scientific underdetermination. In those cases, the phenomena or observations to be explained were the lack of known alternative theories, the occurrence of unexpected explanatory interconnections and the predictive success within the research program. Those phenomena could not be explained by string theory itself or any other physical theory but had to be dealt with at a conceptual meta-level. The one coherent explanation that emerged was the statement on limitations to scientific underdetermination, which then was inferred based on IBE.

In the cases of the final theory arguments, to the contrary, two of the statements that provide the foundation of the argument are implied by string theory itself. The minimal length scale as well as the property of structural uniqueness constitute central elements of the theory. Scientific results in a fundamental scientific discipline like physics, however, are to be expected to attain further explanation, if at all, within the scientific field itself. Searching for philosophical inferences to the best explanation from such scientific results would seem out of place.

So how do final theory arguments work? In the case of the two arguments discussed in Section 6.1, the core reasoning is of a deductive nature. It is claimed that, given that the theory is universal, or, given that a theory enforces a minimal length, certain methods of separating local from global scientific underdetermination no longer work at a conceptual level. An inductive step leads from that analysis to the claim that statements about local underdetermination allow conclusions regarding global underdetermination as well. From there, the argument is then picked up by the three arguments for limitations of local underdetermination, which complete the final theory claim. The overall argument thus contains an element of inference to the best explanation (i.e. the part concerned with local scientific underdetermination), but the specific element usually perceived as the final theory claim does not. The overall argument reaching from the three pieces of evidence analyzed with respect to local underdetermination to the final theory claim based on a statement of strong limitations to global underdetermination then again can be understood as IBE. The fact that there are no possible scientific alternatives to the current theory explains why people found no alternatives, why the theory provides unexpected explanatory interconnections and why effective theories of that theory are explanatorily successful within a certain regime.

The argument from structural uniqueness is of a different nature than the arguments discussed in Section 6.1. It leads directly from a scientific statement (highly predictive structural uniqueness) to the final theory claim. No substantial part of this argument can be phrased in terms of IBE. The argument is a probability assessment relying on a number of individual inductive steps without aiming at explaining the initial phenomenon of highly predictive structural uniqueness. This makes it structurally conspicuously different from the classical IBE-based scheme of scientific reasoning.

6.5 The case of anthropic reasoning

I want to conclude this chapter by returning to a specific case of contemporary reasoning in fundamental physics. The anthropic argument (Susskind, 2003, 2006) has found wide recognition in recent years. Some have hailed it as a fundamental step forward in physical thinking. Others have criticized it as unscientific and detrimental to the longing for genuine scientific progress. It would go beyond the scope of this book to attempt a full assessment of those discussions. The following analysis therefore will not amount to determining the argument's overall status or significance. The anthropic argument is of eminent interest for the core analysis of this book, however, because it signifies the crucial role questions of scientific underdetermination have assumed in contemporary fundamental physics. It is this aspect that shall be discussed in the following.

Anthropic reasoning in contemporary fundamental physics relies on a conjunction of string theoretical arguments and concepts developed in inflationary cosmology. In order to understand the overall argument, we need three conceptual elements: eternal inflation, the string theory landscape and the problem of the fine-tuning of the cosmological constant. We already encountered the multiverse scenario of eternal inflation in Section 4.2; the theory was developed in order to explain basic features of the universe we observe and conjectures a huge, maybe infinite number of universes in an inflationary background space. The string landscape, first mentioned in Section 1.1, denotes the potential of string theory that is understood to have a huge number of local minima each of which corresponds to a specific choice of low energy parameter values. The selection of the actual minimum takes place at the early very hot stages of the universe based on quantum statistics. In a multiverse scenario that selection happens in each universe of the multiverse independently during the transition to the normal expansion phase. This implies that each universe is characterized by its own ground state and therefore by its own selection of low energy parameter values. Given that the number of universes in the multiverse even transcends the number of possible string theory ground states, each individual string theory ground state must be expected to exist in the multiverse.

The anthropic argument aims at using this conceptual framework for solving one of the core problems of cosmology, the problem of the fine-tuning of the cosmological constant. Measurements by Riess *et al.* (1998) and Perlmutter *et al.* (1999) provide strong evidence that the cosmological constant is larger than zero. However, it is about 120 orders of magnitude smaller than one might expect based on general quantum field theoretical calculations of possible vacuum energies. A cosmological constant larger than that fine-tuned value would have generated either an accelerated expansion of the universe that would have led to a cold and mostly empty universe very soon (if the cosmological constant was positive) or (if it was negative) a quickly re-contracting universe that would have ended in a big crunch before ever having allowed for the formations of stars as we know them. A high degree of fine-tuning usually suggests the existence of some mechanism that enforces this fine-tuning. However, despite considerable efforts and a number of specific attempts, no genuinely satisfying mechanism of that kind has been found.

The anthropic argument represents an entirely different approach towards the understanding of fine-tuning. It picks up where we have above left the string theoretical multiverse scenario. We have noted that every possible string theory ground state must be expected to exist in the multiverse. If the fine-tuning of the cosmological constant is roughly of the order 10^{120} and the number of possible ground states is expected to be of a higher order – let us say 10^{500} – we have reason to expect that some ground states with the observed degree of fine-tuning can be built mathematically and therefore also exist in the multiverse. The multiverse thus implies a high probability that fine-tuning is realized in some of its universes on purely statistical grounds without providing a physical mechanism that enforces fine-tuning. On that basis, just one question remains. Why do we live in one of those universes where fine-tuning is the case? Here, the actual anthropic argument kicks in. Since only a fine-tuned cosmological constant provides a framework that allows for the evolution of life (or, let us say, organisms learning and reflecting about the universe), thinking organisms, if they exist at all, must live in one of the fine-tuned universes. The physics thus explains why fine-tuned universes are likely to exist, and the anthropic principle explains why we as thinking organisms sit in one of those fine-tuned universes.

A number of fundamental questions arise with respect to the status of this argument. Some of them have a profoundly philosophical tinge. For example, it is not easy to define the epistemic and ontological status of the other universes in the multiverse. It is clear that they are characterized by observables just like our own. To the extent that they allow the evolution of observers, their properties can be measured from within that universe just like the properties of our own universe can be measured by us. Conceptually, those universes therefore appear as observable physical objects rather than mere mathematical auxiliary constructions. However, one must take into consideration that it is crucial to the anthropic argument that most of the universes do not allow the evolution of observers. In light of this, it seems complicated to speak about observables in a sense that goes beyond the abstract physical definition of observables in the context of a quantum theory. Even if we discard this problem and are willing to talk about per se "observable" phenomena in the other universes of the multiverse, it is clear that these phenomena are unobservable from our own perspective. Moreover, to the best of our knowledge, they are not even causally linked to anything observable to us. The other universes thus belong to a peculiar class of objects which are epistemically inaccessible but have some conceptual characteristics of observable objects.

The situation becomes even more complicated once the anthropic explanation of fine-tuning is introduced. The anthropic argument relies on a statistical ensemble of worlds where one member is observable and the others are epistemically inaccessible to us. It seems difficult to attribute explanatory value to this ensemble without introducing some level of "actual existence" where all those worlds, ours as well as the others, are on par with each other. In other words, the anthropic principle seems to suggest a realist understanding of the multiverse, as a set of individual worlds which exist independently of our perspective. If so, that would make the anthropic argument an interesting example of a physical concept that relies explicitly on a specific philosophical position in the scientific realism debate. (A little more on the question of scientific realism will be said in Chapter 7.)

Be this as it may, I want to focus on a different though related aspect of anthropic reasoning. In conventional scientific contexts, theories which correctly reproduce the empirical data within a certain regime keep their explanatory value within that regime once they have been superseded by a more advanced successor theory. They remain relevant as approximate descriptions capable of explaining the regularities that can be found when observing with a degree of accuracy that justifies the given approximation. Newtonian physics, to state one example, explains the movements of falling apples even in the age of general relativity. This lasting value of physical explanation is an important reason for taking physical explanations seriously even though one must expect current theories to be superseded some day by better and maybe ontologically different successor theories. In the case of anthropic reasoning, we face a substantially different situation. The explanatory value now hinges on the existence of the inaccessible universes. If a successor theory were able to reproduce the cosmological data more precisely without positing other universes, the explanatory value of the anthropic argument would simply vanish. One could not refer to it as an approximately viable explanation because no conceptual trace of those universes would have prevailed in the new theory. The disappearance of the other universes without a conceptual trace can happen because there is no direct empirical trace of those universes that would have to be explained in a successor theory. If we had to expect, however, that the explanatory value of the anthropic argument goes to zero from the perspective of a successor theory, we could not take that argument seriously as a scientific explanation of a physical phenomenon.

This is the point where, once again, the assessment of scientific underdetermination enters the picture. In order to retain explanatory value, the anthropic argument must rely on the assessment that it is highly unlikely that an alternative theory exists which can account for the available cosmological data but does not imply a multiverse structure. This assessment is, of course, a classical case of an assessment of limitations to global scientific underdetermination. An evaluation of the actual value of the anthropic argument in the given context must be based on answering the question as to what extent strict limitations to global scientific underdetermination can in fact be established. The required reasoning does not amount to a fully fledged final theory claim, since successor theories which retain the multiverse structure may not be ruled out. Still, it is a good example for a kind of scientific reasoning where assessments of scientific underdetermination are implicitly built in and constitute a core element of the overall argument. I do not want to analyze the specific nature and strength of the arguments which can be put forward in favor of strong limitations of global scientific underdetermination with respect to eternal inflation. It is important to understand, however, that scientific contexts like the deployment of anthropic reasoning can only be seriously discussed at all with regard to their scientific merits once assessments of scientific underdetermination have been acknowledged as a legitimate part of scientific reasoning.

6.6 A new conception of scientific progress

The final theory claims discussed in this chapter can be analyzed at two levels. First, they appear as a natural continuation, and maybe completion, of the scientific coming of age of assessments of scientific underdetermination described in Part I. Second, however, they introduce a genuinely novel aspect of assessments of scientific underdetermination that reaches out beyond the context of non-empirical theory confirmation. Let me recapitulate both levels of analysis individually.

The central subject of this book is the rise of assessments of scientific underdetermination to the position of an important and irreducible element of scientific reasoning. The case of final theory claims constitutes an important part of this development. As we have seen, final theory claims are particularly strong instances of assessments of scientific underdetermination which are related to arguments of non-empirical theory confirmation in a complex way. Parts I and II of this book introduced assessments of scientific underdetermination as an integral part of scientific reasoning in all of microphysics. The powerful role of non-empirical theory assessment for the assessment of empirically unconfirmed theories in contemporary fundamental physics in this light does not constitute a fundamental break with the past but rather a gradual modification of longstanding methods of theory assessment in physics. The step towards global assessments of scientific underdetermination, to the contrary, constitutes a genuinely new achievement of string physics that has no significant precursors in earlier physics. Global assessments of underdetermination shift the focus from the assessment of a theory's viability within a given empirical framework to the formulation of final theory claims. As we have seen, the "classical" arguments supporting final theory claims in the context of string theory are not self-contained but work merely as indicators for the legitimacy of reducing the question of global scientific underdetermination to the local question. Only in conjunction with the arguments for the limitation of local scientific underdetermination can these arguments be deployed for establishing a final theory claim. The three arguments for local limitations to scientific underdetermination thus remain of substantial importance in the new context.

On the other hand, the final theory claims formally implied by string theory can react back on local assessments of underdetermination and further strengthen the case for non-empirical theory confirmation. The fact that final theory claims are built into the fabric of string theory substantially alters the overall perspective on the scientific process. Once final theory claims turn into genuinely scientific statements, it becomes natural to expect from the scientific process that it can provide means for vindicating those claims. Such a vindication is only possible, however, if theory assessment can meaningfully reach out beyond the horizon of what is empirically testable today. On that basis, it is natural to expect a significant role of non-empirical theory assessment and, eventually, of non-empirical theory confirmation. The fact that a complex web of novel arguments arises in an overall coherent way that connects local and global assessments of scientific underdetermination thus strengthens each individual part of that web of arguments.

The increased significance of local assessments of scientific underdetermination and the theory-driven emergence of final theory claims form a coherent whole.

Section 3.1 has established that local assessments of scientific underdetermination emerge naturally from a conceptually cogent extension of the canonical principle of theory confirmation and constitutes a viable element of critical scientific reasoning. This verdict can now be extended to global assessments of scientific reasoning as well. Since statements on global underdetermination largely rely on assessments of local underdetermination, the examples of scenarios which would weaken local claims of limitations to scientific underdetermination can be directly applied in the global context as well. Any argument that would weaken the status of string theory itself would also weaken the related final theory claims. The third final theory argument, which can be developed independently of questions of local underdetermination, would become really forceful only if the highly predictive character of string theory were established. On the other hand, it would lose its power if the highly predictive character of string theory were conclusively refuted. The overall web of local and global assessments of scientific underdetermination thus allows for the disconfirmation of each of its claims and, on those grounds, can be called genuine scientific reasoning.

Having said that, it is important to keep in mind the considerable difficulties associated with that strategy. The complexity and vagueness of arguments of non-empirical reasoning implies that verdicts will not be as strong as those based on conclusive empirical evidence. The process of non-empirical theory evaluation is much slower and often less consensual than its empirical counterpart. Non-empirical theory assessment requires continuous recalibration in order to minimize the risk of overextending its use and losing the cogent connection between non-empirical reasoning and the principle of empirical testing. For all these reasons, it will always remain the ultimate goal of scientific research to find conclusive empirical evidence in each individual case. The status of a merely non-empirically confirmed theory will always differ from the status of an empirically well-tested one. However, in the light of the arguments presented, this difference in status should not be seen as a wide rigid chasm, but rather as a gap of variable and reducible width depending on the quality of the web of theoretical arguments. One might liken the situation to a comparison between first person and third person observation. While the difference in status is indissoluble, a coherent web of information about some observation by others can make me believe in that observation nearly as strongly as if I had made it with my own eyes. For the growing spectrum of theoretical claims which address aspects of the world where conclusive empirical evidence remains elusive for long periods of time maybe, in some cases, forever, non-empirical theory assessment thus can provide lesser but still significant degrees of confirmation.

The fact that non-empirical theory assessment amounts to *scientific* reasoning marks out the range of the arguments presented in this book. The status of string theory, but also the status of those strategies deployed to assess its status in the absence of empirical data, must prove themselves by being confronted with observations and scientific analysis – both at the level of the theory's predictions and at the meta-level described in this book. These assessments have to be made on scientific rather than philosophical grounds. The analysis provided by this book therefore cannot establish the actual power of the presented strategies of assessing scientific underdetermination, let alone the viability of string theory or any other scientific theory. It can only establish that these strategies are actually applied in science, that they are rationally sound in principle and that they can be powerful under certain external conditions. To what extent those conditions apply depends on the world we face. And that world, in the end, has to be learned about by observing it.

6.7 The end of science?

Claims of a final theory smack suspiciously of a declaration of the end of science. The current condition of string theory, however, suggests a quite different conclusion. In order to understand the peculiar character of the situation in string theory today, we have to compare it once again with more traditional physical research.

The standard conception of scientific progress takes for granted a significant disparity between experimental search and theoretical fit with respect to their respective time horizon. While the discovery of new physical phenomena is elevated to an eternally valid principle of scientific research that puts a final empirical inventory (and consequently a final physical theory) forever beyond the grasp of human inquiry, the completion of the specific scientific theories which fit a certain set of phenomenological data is considered a finite and predictable enterprise. The scientists who take on the challenge to create a theory about some phenomena based on some set of principles and assumptions are expected to be able to complete that theory as a coherent and calculable structure (modulo some minor unsolved aspects, perhaps) within a reasonable amount of time.

The distinction between the infinite duration of the quest for new physics and the limited creational period of single theories clearly mirrors the reality of traditional scientific research. In the case of string theory, it seems oddly out of place. If the final theory claim associated with string theory is correct, the time until we reach a final description of nature is reduced to the time we need to complete string theory itself. This step, however, goes along with a vastly extended time horizon for the completion of this one theoretical scheme.

It may be helpful at this point to remember the situation four centuries ago, when philosophers like Francis Bacon and René Descartes laid the foundations of the scientific world-view. Bacon as well as Descartes considered the creation of the scientific method itself to be the main achievement on the way towards a correct understanding of the world. The scientific details in their understanding could then be filled in within a lifetime. History has confirmed the immense fertility of the scientific method, but it has also disappointed the early expectations of imminent full enlightenment. The scientific method turned out to be a starting point for many generations of ever deepening research, which, despite its success, did not reach the elusive endpoint of a full description of nature.

The situation in string theory today might bear some resemblance to the early history of science. String theory, by establishing the notion of a final theory, might well prove to be of great significance for our understanding of nature, but might nevertheless disappoint hopes to provide a specific time frame for the fulfilment of its ultimate promise. We are in no position today to assess whether – and, if so, when – a full formulation of string theory will be found. At any rate, we have no reason to assume that it will happen anytime soon. The arguments by analogy derived from the examples of earlier theoretical schemes fail because of the significantly different level of complexity of string theory and its nature as a final theory. One might well compare the seemingly never-ending sequence of theoretical problems arising on the way towards a complete understanding of string theory with the sequence of new physical phenomena that characterizes traditional physical research. In the same way as the perspective has emerged in the traditional scientific setting that science will always face new phenomena and, therefore, will never reach a final theory, one could now be led to suspect that string theory will always face new theoretical problems and will therefore never become fully mature. String theory thus should not be taken to announce an end of science but rather to represent a new phase of scientific progress. In this new phase, progress in fundamental physics is no longer carried by a sequence of limited, internally fully developed theories, but rather by the discovery of new aspects of one overall theoretical scheme whose general characteristics identify it as a candidate for a final theory, yet whose enormous complexity bars any hope of a full understanding in the foreseeable future.

7

An altered perspective on scientific realism

7.1 The layout of the debate

Up to this point, the analysis of this book has largely been confined to the scientific assessment of physical theories. It was concerned with the way scientists evaluated the chances that their theories were empirically viable or constituted final theories. This chapter will turn to a more far-reaching philosophical question: the question of scientific realism. The scientific realism debate addresses a pivotal question of scientific reasoning. How and in what sense does science connect to the world? Do our scientific theories tell us something true about reality that goes beyond the mere structuring of observations and the predictions of future empirical data? Scientific realists insist that they do while scientific anti-realists deny it. As it turns out, the analysis carried out in previous chapters has significant implications for the question of scientific realism. In order to appreciate the philosophical role of assessments of scientific underdetermination in this context, it is helpful to start by choosing a wider perspective, however. Before addressing the question of scientific realism directly, let me briefly look at a related elementary epistemic debate, the debate about rationalism.

The old debate between rationalism and empiricism revolves around the nature of our access to knowledge about the world. Broadly construed, rationalism asserts that the most profound or most stable knowledge about the world can be generated by pure rational reasoning. Early examples of rationalist thinking in this sense are the so-called proofs of god developed by scholastic philosophers. Rationalism flourished in the seventeenth century under philosophers like Leibniz and Descartes. Kant, though rejecting the formers' philosophies, still retained traces of rationalism by rationally deducing the a priori status of a specific class of synthetic statements which he considered empirically irrefutable. Speculative philosophers of the nineteenth century like Hegel once again endorsed a more far-reaching belief in rationalist argumentation.

157

The opposite position of empiricism insists that all our knowledge about the external world must be based on empirical observation. This position is often softened in two ways. First, it is held that logical and mathematical knowledge does not refer to the external world and therefore can be acquired without observation. Second, it is conceded that inductive reasoning can legitimately lead to generalizations from actual observations. In this understanding, rational analysis can be deployed in order to structure, systematize and generalize our observations. It cannot, however, lead to knowledge about kinds of things which we have not observed.

The emergence of modern science was based on developing a way of reasoning that aimed at supporting all its statements by thorough observation and experimental data. The striking success of the scientific method thus propelled an empiricist perspective on human knowledge to a highly influential position in philosophical thinking. Most exponents of science as well as most philosophers with sympathies for scientific reasoning took the decision to abstain from empirically unfounded rationalist speculation to be a core element of the success of science and, on that basis, adopted a generally skeptical attitude towards rationalist speculation. Moreover, scientific progress repeatedly refuted claims which had been established by pure philosophy on the basis of pure rational reasoning. Famously, the development of non-Euclidian geometry and its later deployment in general relativity discredited Kant's assertion of the synthetic a priori character of Euclidian space. Notoriously, the discovery of Neptune ridiculed Hegel's philosophically derived statement that the solar system must have seven planets. Instances of that kind contributed to the emerging understanding among scientists that rationalism constituted a counterproductive way of thought that had to be kept out of any serious analysis in order to avoid wrong or nonsensical conclusions. Philosophical approaches which aimed at accounting for the success of science largely followed that assessment. The philosophy of science, in particular, was bound to remain close to the principle of the epistemic primacy of empirical data that carried the scientific enterprise itself.

Once rationalism is rejected and the primacy of empirical data is acknowledged, the philosophical question of scientific realism arises. Given that scientific statements can be informed and confirmed only by empirical data, can they constitute knowledge about anything beyond observational data at all? Scientific anti-realists follow the empiricist line of reasoning all the way through and give a negative answer to that question. They claim that scientific theories cannot establish knowledge about the existence of scientific objects in the external world. Knowledge, according to their understanding, is exclusively about the phenomenal data presented to us. The scientific realist, to the contrary, wants to retain at least some trace of the pre-scientific autarky of conceptual thinking. She awards to unobservable scientific objects like the electron or the quark the same ontological and, if sufficiently well confirmed, the same epistemic status as to chairs and tables.

The question of scientific realism rose to prominence in the nineteenth century in the context of the atomism debate that was addressed in Section 5.2 of this book. Two related developments pulled in different directions at the time. On the one hand, abstract science had been developed to a sufficient degree to formulate scientific empiricism in its fully fledged form. Physics and chemistry could be understood as entirely formalized methods for grasping the regularities observed in nature and extracting predictions of future empirical data on that basis. Abstract concepts like energy conservation played an important role in that kind of formalized scientific analysis and seemed to suggest that science was moving ever farther away from the "primitive" physics that relied on our intuitions about observable objects. On the other hand, during the nineteenth century the old idea of atomism, which was based on such a "primitive" intuition, began to find genuine and significant empirical support in physics and chemistry. The questions at the time were (i) whether one could conclude from the agreements between empirical data and implications of the atomist hypothesis that the deployment of the atomist hypothesis would be conducive to scientific progress in the future and (ii) whether one could conclude from these agreements that the hypothesis was true.

Ignoring intermediate positions, two opposing core positions could be identified. The realist atomist held that the clear correspondence between the predictions of atomism and the available empirical data had to be understood as a clear indication that atoms existed, even though they remained unobservable. If atoms existed, however, one of course had to continue using the concept in future scientific research. The anti-atomists argued that any attempt to introduce intuition-based concepts in abstract physical descriptions of the world constituted a step backwards from the modern scientific methodology, which consisted in the pure and straightforward adaptation of mathematically formalized theories to the known empirical data. Understood in this way, science did not offer any basis for declaring the truth of "metaphysical" speculations about the existence of unobservable objects. Such speculations might for a while cohere with the path of actual scientific progress but had to be expected to become inadequate at some later stage. For that reason, belief in the truth of those speculations was not just erroneous but also scientifically counterproductive by providing a metaphysical excuse for sticking to empirically inadequate concepts in the future. In a nutshell, the atomist/realist argued that science had led to a situation where it was utterly irrational to overlook the empirical evidence for the existence of atoms while the anti-atomist/empiricist argued that the structure of scientific reasoning offered no conceptual basis for acknowledging the existence of atoms *as a matter of principle.*

While the atomism debate in the nineteenth century was a genuinely scientific debate on the choice of the most adequate research strategy, the scientific discoveries of the early twentieth century – which were briefly sketched in Section 5.2 – solved the scientific dispute and shifted the debate entirely to the philosophical level. Regarding question (i) above, physicists came to agree that the atomic concept as well as other concepts about microphysical objects were productive as long as their properties were adapted freely and often counter-intuitively to the empirical data. Question (ii) was left to the philosophers, who debate it to this day.

It is impossible to present the multifaceted debate on scientific realism that has evolved in recent decades to any degree of detail in the present context. I just want to lay out a few basic ideas which will play a core role in the ensuing analysis.

The scientific realist has two basic worries about a strictly empiricist understanding of science. The first one is an intuitive worry. The empiricist understanding of science just does not ring true. It does not seem to cohere with the way scientists think and behave. Scientists attribute considerable significance to the discovery of new microphysical objects. Conspicuously many Nobel prizes in physics, to give an example, have been awarded for discoveries of particles rather than for the confirmation of abstract physical principles. Once microphysical objects have been confirmed empirically, physicists tend to have the same kind of trust in their behavior as in the behavior of observable objects. In many cases, physicists even use micro-physical objects as material tools for manipulating and testing other micro-physical setups. (This aspect was emphasized by Hacking, 1983). In cases like cells, we can see the objects through a microscope, know a huge amount of detail about them and have a picture that is so close to the picture we have of complex observable objects that we might actually at times forget the cell's unobservable character. It seems more than a bit strange to ignore all this and stick to the idea that we have no cognitive access at all to the question whether the involved scientific objects actually exist.

The second realist worry, which is of a more technical nature, is often formulated in terms of the no miracles argument (NMA). In a nutshell, NMA asserts that scientific realism is the only position that can explain the frequent predictive success of science and should be adopted for that reason. NMA was first formulated by Hilary Putnam in a brief paragraph in Putnam (1975) and soon developed into the most popular and most intensely discussed argument for scientific realism. NMA is a three-step argument. First, it is asserted that the frequent predictive success of science looks like a miracle as long as one does not assume scientific realism. Then it is argued that scientific realism can in fact provide a satisfactory explanation of the predictive success of science. Finally, inference to the best explanation leads to the conclusion that scientific realism is probably true. Some philosophers (see e.g. Musgrave, 1985) have specified step one by emphasizing that only successful predictions of genuinely novel phenomena require a realist explanation, while the frequent occurrence of correct scientific predictions which constitute mere extrapolations of a pattern of observations are explicable based on the validity of the principle of induction without any further assumptions. This more specific understanding of NMA shall be adopted in the discussion below. One important aspect of NMA, which was particularly emphasized by Richard Boyd (1984, 1990), is the fact that NMA constitutes an empirics-based argument. It relies on the *observation* that science is often predictively successful. Boyd's point already indicates what will emerge more clearly in the course of this chapter: NMA is related to the kinds of reasoning that have been discussed in this book in the context of non-empirical theory assessment.

The scientific realist asserts that realism is the correct conclusion to be drawn from both arguments sketched above: realism can account for our intuitive understanding of the research process as well as for the phenomenon of successful predictions in science. But what precisely does scientific realism consist of? One may identify two interconnected elements which, in conjunction, amount to genuine scientific realism.

First, the scientific realist must assume that there is a deeper level of reality underlying the phenomenal level of empirical data. That level, the realist asserts, can be addressed by scientific theories. Scientific statements addressing that level tell something substantially different than corresponding statements about mere regularities in the observed data. To give an example, the statement that electrons exist and behave in a certain way is assumed to tell something substantially different than the set of statements which describe all regularities in empirical data which arise due to electrons. Scientific realists have different opinions as to what kind of deeper-level statements the realist should focus on. For example, ontological realists speak of real objects while structural realists speak of real structures. What scientific realists agree upon, however, is the claim that an underlying reality of our observations can be identified at some level and is of substantial importance.

Already this first claim is questioned by rigid forms of empiricism. Logical empiricism, the position that was predominant in the philosophy of science from the 1930s to the 1960s, deployed a theory of meaning that was so narrowly based on empirical confirmation that it did not allow for any meaning differences between statements on physical objects and the corresponding phenomenal statements. Logical empiricists thus denied already the first claim of scientific realism and took scientific realism (in the sense presented above) to be meaningless on that ground.[1] Current conceptions of empiricism, in particular Bas van Fraassen's constructive empiricism and empirical stance (van Fraassen, 1980, 2002), endorse a less rigid theory of meaning and therefore are ready to grant to the scientific realist that, by talking about scientific objects, she is indeed talking about something different than the observable implications of those objects. In van Fraassen's words, scientific statements about unobservable objects can be "literally true." Modern empiricists therefore mostly have no principled objection against the realist's first assertion. What they do problematize, however, as we will see in the following, is that the scientific realist fails to offer a clear specification of the level at which she wants scientific statements to be understood.

The second claim of scientific realism asserts that well-established scientific statements which address the described level of actual scientific objects, structures or anything else are mostly approximately true. For example, the ontological realist believes that the objects posited in well-established scientific theories mostly refer to things in the external world which have properties that are grasped approximately by the theory in question. It is this second claim that is rejected by modern empiricists. They deny that we can establish the literal approximate truth of our best scientific statements. In addition, they also deny that we have a good reason for assuming that scientists aim at establishing the approximate truth of scientific statements. The empiricist presents a wide spectrum of arguments against a realist understanding of scientific objects.

The first anti-realist argument is not directed against the coherence of scientific realism but rather against the main motivation for scientific realism, NMA. It is denied by many anti-realists that the prognostic success of science, which constitutes the point of departure for NMA, requires any explanation that reaches beyond a simple analysis of the research process itself. Empiricists often hold that the impression of a conspicuous sequence of predictive success in science is an artefact produced by a historical perspective that focusses on those theories that have turned out successful and neglects the mass of scientific ideas that have led nowhere (Fine, 1986). A similar argument at a slightly

---

[1] In their own terms, logical empiricists would understand realism as a mere choice of language that uses the concept of real objects but does not imply any philosophically significant differences to phenomenalism (that is, the choice of a phenomenal language.)

different conceptual level suggests that the high degree of accordance with empirical data to be found in well-established scientific theories is based on a kind of Darwinian selection mechanism that lets all those theories die which were not up to the same standard (van Fraassen, 1980).

The second anti-realist argument is directed against the argumentative value of scientific realism. Empiricists argue that, even if the phenomenon of the predictive success of science indeed left something to be explained, the realist would be scarcely better off than the empiricist (see e.g. Fine, 1986). The realist's problem, according to that line of criticism, is that she cannot believe in the absolute empirical adequacy of present scientific theories. It is scientific consensus in next to all scientific contexts that our present theories are bound to be superseded by empirically more accurate ones at some stage.[2] Thus, the scientific realist must settle with approximate empirical adequacy of present theories, which implies that today's theories can be approximately true at best. If so, however, the degree to which a present theory is true cannot be known to imply that the theory is in agreement with future empirical data and will turn out predictively successful. Approximate truth would at each stage be fully compatible with the theory's disagreement with future data since that future data could happen to rely on those of the theory's elements which are not true. On that basis, however, it seems unclear in which sense scientific realism can be a good explanation of predictive success. If approximate truth does not imply predictive success, how should it explain predictive success? One might still try to argue that the theory's approximate truth implies a higher probability of predictive success than mere approximate empirical adequacy. This claim, however, is by no means self-evident and has turned out difficult to support along a genuinely realist line of reasoning.

A third problem raised by the empiricist has to do with the historical record of scientific theory building. As Larry Laudan (1981) points out, there are many examples of theories which were scientifically and even predictively successful for a long period of time but eventually had to be superseded by new theories with an entirely different ontology. If the ontology of the superseded theory does not survive, however, a classical (that is, ontological) scientific realist cannot call them approximately true. Calling a theory with a completely false ontology true would seem to deflate a theory's (approximate) truth to its (approximate) empirical adequacy, which would amount to dissolving the distinction between realism and empiricism altogether. (An attempt to retain a distinction without relying on ontological objects is structural realism, which will play an important role in Section 7.4.)

---

[2] The only exception, of course, is string theory where a final theory claim has been formulated.

Laudan's observation can be applied at two different levels. The stronger conclusion is called the pessimistic meta-induction. It is based on the understanding that examples of fundamental shifts of ontology despite strong empirical success of the superseded theory are so common in science that they must be taken as a characteristic of the historical evolution of science up to this day. If that is the case, however, we would need a strong reason to deny that we have to expect similar ontological shifts also with respect to the theories we endorse today. If, as the supporter of the pessimistic meta-induction believes, no strong reasons of that kind are in sight, we cannot hold our present theories approximately true. Even without believing that fundamental shifts in ontology after considerable predictive success are common in mature science, however, the fact that they occur at all constitutes a serious problem for NMA. In cases where they occur, it seems that we have to find an explanation other than scientific realism for the success of the theories in question. If an alternative explanation does exist in those cases, however, it seems difficult to argue that it cannot possibly apply in other cases as well. Therefore, step one of NMA, which amounts to the claim that there seem to be no other explanations of strong predictive success than scientific realism, is in serious danger.

7.2 Unconceived alternatives

There is one argument that grasps with particular clarity the general conceptual impasse scientific realism finds itself in. This argument of transient underdetermination, which was first formulated by Lawrence Sklar (1975, 1981) and recently further elaborated by Kyle Stanford (2001, 2006), will eventually lead back to the line of reasoning followed in this book.

The approach chosen by Sklar and Stanford is of particular interest from the perspective of this book since it relies on the concept of transient underdetermination. As discussed in Chapter 2, transient underdetermination by and large resembles the concept of scientific underdetermination that has played a crucial role in the analysis of this book. Sklar and Stanford make the point that the claim of scientific realism amounts to a statement on unconceived alternatives.[3] If the realist learned about the existence of scientifically equally satisfactory but ontologically – or structurally – different unconceived alternatives to the present theory, her conviction that the objects or structures of the present theory refer to something in the external world would be substantially reduced. A strong claim that the objects or structures of a theory do refer thus must be based on the

---

[3] This is the term used by Stanford.

assumption that unconceived alternatives of the mentioned kind are unlikely to exist. Sklar doubts, at any rate with respect to fundamental physical theories, that science has the means for assessing unconceived alternatives and thus for addressing the question of transient underdetermination. If scientific realists claim to be able to extract realist conclusions from scientific reasoning, they therefore disregard the limitations of genuine scientific reasoning and require a service from the scientific process which the latter is probably incapable of delivering. Stanford further strengthens Sklar's point (and extends its range to contexts beyond fundamental physics) by emphasizing that there are many examples of theories which had been taken to be without alternatives at some stage while alternatives were in fact developed later on, which rendered the earlier assessment manifestly incorrect.

Sklar and Stanford present their argument pragmatically. They take it to be a matter of the observed characteristics of scientific research whether or not assessments regarding unconceived alternatives are carried out and can be trusted in a scientific field. While they are skeptical regarding the value of assessments of unconceived alternatives to theories about fundamental constituents of matter, they do not deny that assessments of the probability of unconceived alternatives can in principle enter scientific reasoning. Sklar suggests that assessments of the probability of unconceived alternatives may be possible in cases of theories in special sciences which address a comparably narrow spectrum of phenomena.

Still, the perspective of unconceived alternatives presented by Sklar and Stanford can be used for providing a particularly clear account of the inherent tension between scientific realism and the canonical understanding of the scientific process. According to the latter understanding, scientific reasoning is concerned with theories that have been developed in order to account for the data that has been collected. Scientific theory confirmation on that account is constrained to the evaluation of the theory's capability of reproducing the available data and making successful predictions. All other criteria of theory evaluation are "soft" criteria which remain unscientifically subjective and therefore can never amount to scientific knowledge (see Chapter 2). Sklar and Stanford now point out that scientific realism, if analyzed carefully, is concerned with an entirely different question than "canonical" empirical theory confirmation. A statement of scientific realism is not primarily based on the assessment of the theory that has been developed but first and foremost on an assessment of the spectrum of unconceived alternatives or, to put it differently, of the limitations to scientific (in the words of Sklar and Stanford, transient) underdetermination. But the canonical strategies of scientific theory assessment simply don't reach there. The statement that a given theory has no alternatives cannot be deduced from that theory's scientific predictions. Confirming those predictions thus cannot confirm the statement that there are no alternatives. It follows that the assessment of unconceived alternatives cannot be a genuine part of scientific reasoning. Claims of scientific realism, which rely on assessments of unconceived alternatives, thus cannot be legitimated by scientific reasoning either. The argument of unconceived alternatives provides a convincing demonstration that a modern form of empiricism follows naturally from the canonical paradigm of scientific theory assessment.

Scientific realists have tried to counter the argument of unconceived alternatives by showing that assessments of the probability of unconceived alternatives are in fact carried out in scientific reasoning about unobservable objects and can be reliable in that context. A recent example of that line of reasoning is Roush (2006). Roush argues that, while it is true that *some* contexts of empirical confirmation of theories which posit unobservable objects offer no reasons for assuming that the confirmed theory has no unconceived alternatives, other contexts do provide such reasons. Her specific example has already made its appearance in this book – the confirmation of atomism by Perrin in the early twentieth century. Roush focusses on Perrin's measurement of Brownian motion[4] rather than, as is more common, on the coherent system of his various measurements of the Avogadro number. She does so for a specific reason: she wants to establish that there can be very simple hypotheses in fundamental physics which allow for equally simple arguments against alternatives to these hypotheses. Simplicity, in her eyes, is the basis for keeping the space of possibilities under control and therefore for allowing claims about the probability of unconceived alternatives. In the case of the atomist conjecture in the context of Brownian motion, the core atomist claim just asserts that there are microphysical objects which behave roughly according to the laws of classical mechanics and cause Brownian motion of observed objects by colliding with them. Roush now makes the following claims. First, any alternative claim to the simple form of atomism involved in the given explanation of Brownian motion must include the statement that there are no nearly classical microphysical objects in the system. Second, the mere observation that the movements of the test particles are thoroughly random implies for all practical purposes that these movements are caused by statistical collisions of the test particles with microphysical particles. In other words, Roush argues that the system of dichotomies is simple enough to allow a virtual exclusion of unconceived alternatives based on the observation that the motion of Brownian

---

[4] Brownian motion is an effect that was first observed in the early nineteenth century. Minuscule particles suspended in a liquid were observed under a microscope to move along seemingly random irregular paths.

particles is genuinely random. She thus concludes that, by demonstrating the latter point, Perrin conclusively established the existence of atoms.

Stanford (2009) has answered Roush by pointing out that her construal of a cogent connection between random movements and molecular collisions is based on a rather narrow framework of presumptions which didn't have to be shared by critics of atomism in the early twentieth century. For example, Roush's reasoning relies on the validity of determinism. Giving up the assumption that very small particles move through free space according to deterministic laws would have been an entirely legitimate scientific suggestion at the time, however. After all, something similar was indeed suggested later on in the context of quantum physics.[5] A suggestion of the described kind could have provided a basis for explaining the random movement of test particles based on a stochastic law of movement (e.g. under the influence of electromagnetic force) without relying on atomism. Stanford rightly concludes that the exclusion of unconceived alternatives in Roush's scheme is an artefact of the choice of a specific scientific framework that was itself not without alternatives.

I suggest that Roush's approach runs into the described problems because it does not emancipate itself sufficiently from the canonical paradigm of theory confirmation. Roush assumes that statements about the probability of unconceived alternatives can be tested along the lines of the canonical understanding of theory confirmation. She claims that, if a hypothesis and the methods of its empirical confirmation are simple enough, we can exhaustively understand the space of possibilities and thus infer "semi-deductively"[6] that no alternative hypotheses are possible which agree with the data confirming the given hypothesis. The very same experimental process that confirms the scientific hypothesis in question therefore is taken to virtually exclude all possible alternatives as well. The idea that a "semi-deductive" step leads from empirical theory confirmation to the exclusion of unconceived alternatives remains dangerously close to rationalist thinking, however, and fails for the same reasons as other variations of rationalism: it seeks deduction where no deduction is to be had. Once one chooses a wider perspective on the spectrum of possible scientific structures, the cogency of the semi-deductive step disappears.

---

[5] Ideas of indeterminism had in fact been discussed already since the late nineteenth century (see Stöltzner, 1999).

[6] I call this step "semi-deductive" since it is reminiscent of Newton's idea of the "deduction from the phenomena" (see footnote 26 in Section 1.3). The debate between Roush and Stanford shows some similarities to the debate between Norton and Worrall on the status of "deduction from the phenomena" (Norton, 1993, 1994; Worrall, 2000).

The approach of limitations to scientific underdetermination presented in this book concurs with Roush in claiming that assessments of unconceived alternatives do play an important role in the scientific process and constitute an irreducible element of theory evaluation in physics and other scientific fields. But it suggests a very different mechanism that makes these assessments possible: assessments of scientific underdetermination are taken to be based on empirical testing and inductive inference at a meta-level. The suggested mechanism thus does the very opposite of what is argued for by Roush. Rather than narrowing down the framework of the analysis, it extends it by including all explanatory interconnections provided by the given theory and by taking into consideration the empirical successes of all other theories in the field. With respect to Perrin, it thus vindicates, as discussed in Section 5.2, the traditional understanding that it was the web of coherent arguments for atomism rather than one semi-deductive strand of reasoning that established atomism as a confirmed hypothesis.

The suggested approach to assessments of limitations to scientific underdetermination sidesteps the problems of Roush's suggestion because, by connecting to empirical testing at a meta-level, it avoids unsustainable traces of rationalist reasoning. Unlike Roush's approach, however, it does not, on its own, imply scientific realism. In order to understand that a theory is approximately true, we must have an opinion on the overall number of unconceived alternatives, not just on those which are distinguishable by the next generations of experiments. That is, we must address the question of global scientific underdetermination. As discussed in Section 3.3, the three widely applicable strategies of assessing limitations to scientific underdetermination (that is, NAA, UEA and MIA) all merely address the question of local scientific underdetermination. The verdict of Sklar and Stanford that we have no understanding of the overall number of unconceived alternatives thus is not directly threatened by assessments based on those three strategies.

Still, the realist philosopher may find an improved basis for arguing her case. Contrary to the assumption of Sklar and Stanford, even scientists in fundamental science frequently resort to assessments of the number of unconceived alternatives in their everyday reasoning. In fact, assessments of unconceived alternatives seem most powerful and influential in the context of the most fundamental theories in physics. They are deployed in those contexts in order to develop trust in empirically unconfirmed theories as well as in the novel predictions of theories which have already been empirically confirmed. The philosopher therefore charts known scientific territory when resorting to similar assessments in her argument for scientific realism. The genuinely philosophical element of reasoning in this light just consists of raising the level of discussion from the level of local assessment to the level of global assessment of scientific underdetermination. And here the realist philosopher may resort to her own argumentative strategy along the lines of the no miracles argument. Though local assessments of scientific underdetermination do not establish scientific realism, they thus may be taken to make philosophical arguments for scientific realism look more plausible from a scientific perspective.

If we look specifically at string theory and take into account the final theory arguments discussed in Chapter 6, the support for the realist case against the argument of unconceived alternatives gets decidedly stronger and more direct. With final theory claims, global assessments of scientific underdetermination become part of scientific reasoning in physics. Using the physical final theory claims, the philosopher now can directly block the argument of unconceived alternatives against scientific realism. Nevertheless, even a well-supported final theory claim does not fully establish scientific realism as it was defined in Section 7.1. A crucial element is still missing: no specification has been given of a non-phenomenal level of description at which the literal truth of the final theory can be asserted.[7] The specific analysis of the implications of final theory claims for the question of scientific realism shall be continued in Section 7.5. At the present stage, however, it is important to make a more general point: assessments of scientific underdetermination are by no means univocally favorable to scientific realism. The following section will show that they actually threaten the core argument for scientific realism, namely NMA.

7.3 Scientific realism and non-empirical theory evaluation

It was pointed out above that local assessments of scientific underdetermination do not amount to scientific realism. Nevertheless, as was argued in Part I of this book, they can explain the predictive success of science: repeated predictive success arises because coherent scientific theories are a scarce good. Once scientists have found one, they can be confident under certain conditions that few or no alternative theories exist which are empirically distinguishable at the next stages of empirical testing. This assessment implies that the theory they know has a decent chance of predictive success. The described explanation of predictive success creates serious trouble for NMA, the core argument for scientific realism. As shown in Section 7.1, NMA is a three-step argument.

---

[7] On the other hand, it should be noted that the final theory claim is stronger than a claim of scientific realism in an important respect. Theory succession is not excluded by the realist as long as it does not topple the ontology or structure associated with the present theory.

The first step is the assertion that there are no other explanations of predictive scientific success than scientific realism. Now, since local limitations to scientific underdetermination do not amount to scientific realism themselves, they constitute an alternative explanation of predictive success in science. If such an alternative explanation based on limitations to underdetermination works, however, realism cannot be called the only available explanation of predictive success in science. Therefore, step one of NMA fails.

To make matters worse for the scientific realist, limitations to scientific underdetermination may actually be a better explanation of predictive success than scientific realism. In several respects, they avoid criticism faced by the latter. Let us first look at Laudan's pessimistic meta-induction, which asserts that the approximate truth of posits of unobservable ontological objects is contradicted by the history of past ontological changes. Scientific realism is vulnerable to this argument because it is based on the absolute statement that the ontology of our current theory will survive until the end of science. The discovery that one single substantial change of the present theory's ontology must be expected to occur therefore is sufficient for confuting the realist claim with respect to that theory. The claim of limited scientific underdetermination, to the contrary, is compatible both with the occasional failure of predictions and the occasional substantial change of ontologies. In order to refute a claim of limitations to scientific underdetermination, one would have to show that no significant pattern of empirical predictive success can be found in the scientific field under investigation. Laudan's argument, however, does not aim at providing anything of that kind and therefore cannot threaten the claim of limitations to scientific underdetermination.

As pointed out in Section 7.1, Laudan's reasoning contains a second, less far-reaching conclusion from the observation of ontological breaks in the evolution of science. Even a single instance of a predictively successful theory that eventually is replaced by an ontologically substantially different one threatens NMA because it provides an example of a case where predictive success must have a different explanation than scientific realism. NMA asserts that predictive success allows a conclusive inference to the approximate truth of the corresponding theory. If examples of predictive successes provided by false theories can be found, NMA thus fails. The argument of limitations to scientific underdetermination remains unaffected by this second claim of Laudan as well because it makes less risky claims than scientific realism in two respects. First, it does not imply that theories which have been scientifically successful must be approximately true but merely argues that scientific underdetermination must have been limited in that case. Second, the meta-inductive inference from limitations to scientific underdetermination in one context to such limitations in a context to be tested in the future may not be strict either. It merely constitutes a general judgement that limitations to scientific underdetermination tend to be strong in the field. This does not exclude the possibility that a specific theoretical question allows for many solutions and therefore provides no basis for trustworthy predictions even though predictions were possible at an earlier stage of the theoretical evolution. What the defender of a principle of limitations to scientific underdetermination must hold is merely that the scientists, once they have understood the theoretical framework sufficiently well, are usually capable of making an assessment as to whether or not a case for limitations to scientific underdetermination can be made. The explanation of predictive success by limitations to scientific underdetermination thus is far more flexible than a realist explanation in dealing with altered situations in the course of scientific progress.

The analysis of assessments of scientific underdetermination also seems favorable to canonical scientific realism in another respect: it provides a qualified counter-argument against the assertion – mentioned in Section 7.1 – that predictive success does not occur with significant regularity in science. The general status of scientific predictive success may be characterized in the following way. Though many scientists take a substantial predictive success rate of modern scientific theories as a self-evident fact, it is not possible to extract an objective numerical success rate of science that could conclusively refute the hypothesis that the notion of significant predictive success in science is merely created by a biased historical perspective. While the kind of reasoning presented in this book does not allow the extraction of a numerical success rate of science either, it provides a new perspective from which a denial of significant predictive success in science becomes less plausible. The presented analysis has led to the conclusion that the predictive success of scientific theories plays a crucial role not just in the assessment of empirically unconfirmed theories but also in the assessment of theories which, according to a modern understanding of theory confirmation, count as empirically confirmed. It was argued in Sections 5.2 and 5.3 that the conception of theory confirmation in modern microphysics relies on assessments of scientific underdetermination which, in turn, are based on the experience of regular predictive success within the research program. In other words, without the understanding that there is regular predictive success in science the modern concept of empirical confirmation could not be justified; and if that understanding were fundamentally flawed, the modern concept of theory confirmation would so often have failed to be reliable that it presumably would have never been established.

Let me explicate this point by looking once more at an example from particle physics. The understanding that the observation of a particle in a cloud chamber constitutes empirical confirmation of that particle implies that one can expect to find effects of that kind of particle in other contexts which are in agreement with the theories that have been established at this point. These effects, to the extent they have not been observed yet, constitute novel predictions of the theory. The notion of confirmation of that particle only makes sense if those predictions tend to be fulfilled. Otherwise, one would continually be dealing with concepts which theoretically have far-reaching implications that empirically never materialize. A scientific environment that does not allow for optimism regarding predictions related to confirmed objects does not support the notion of the empirical confirmation of physical objects at all. Empirical confirmation would have to retreat closer to the phenomenological surface where predictions have a more immediate connection to enumerative induction. The fact that the confirmation of scientific objects plays a crucial role in particle physics thus must be counted as strong evidence for the claim that a significant rate of predictive success does occur in that field.

It now becomes clear why empiricists are skeptical with regard to significant rates of predictive success: they ask the wrong question. The plausibility of the skeptic claim is based to a considerable degree on its generality. If we look at all scientific theories that were ever developed, it is probably true that most of them turned out false. It seems equally true, however, that most of them were never given particularly high chances of success by the scientific community. The important role of assessments of non-empirical theory assessment in science suggests that the empiricist who denies significant rates of scientific success avoids asking the question that is most important for understanding scientific success. That question is not: why does science mostly produce predictively successful theories? It rather is: why are scientific theories mostly empirically successful when scientists expect them to be successful? The second characterization of predictive success in science is far less dubitable than the first. It is this second characterization of significant rates of predictive success that is addressed by explanations based on the assessment of limitations to scientific underdetermination. The claim of limitations to scientific underdetermination does not amount to the statement that underdetermination is strongly limited in all or most questions which arise in science. Rather it makes two different assertions. First, it states that strong limitations to underdetermination occur frequently; and second, it states that scientists have fairly reliable – though by no means infallible – tools to understand when strong limitations to scientific underdetermination are present.

To conclude, local limitations to scientific underdetermination in a number of respects seem to constitute a more satisfactory explanation of predictive success in science than scientific realism. Therefore, they seem to remove the predictive success of science as a convincing argument for scientific realism.

Sections 7.2 and 7.3 demonstrate that the acknowledgement of an important role of assessments of limitations to scientific underdetermination in the scientific process offers mixed signals regarding scientific realism. While it weakens core empiricist arguments against scientific realism, it also undermines the case for scientific realism by demonstrating that the tasks often understood to be performed by scientific realism may also be accomplished by other means. This ambivalent message can be transformed into a more coherent statement when formulated in terms of a comment on the dichotomy between rationalism and empiricism. The construal of the latter dichotomy was based on the understanding that the reliance on empirical confirmation, that is, on confirmation by empirical data of the kind predictable by the theory to be confirmed, constituted the main scientific counter-concept against rationalist reasoning. Any kind of reasoning that did not adhere to this principle thus immediately came under suspicion of amounting to a relapse to rationalism. The empiricist consequently took the modern observer of the world to face the following choice: either to abstain from any far-reaching claims on the status of scientific theories and feel content with the scientific successes of saving the phenomena or to relapse into discredited and futile rationalist thinking. The understanding that statements of scientific realism cannot be justified by scientific reasoning followed from this.

The analysis presented in this book suggests that the empiricist's understanding is inadequate. Parts I and II of the book demonstrate that modern scientific reasoning includes an important element of non-empirical theory assessment that contradicts the empiricist conviction that scientific theory assessment cannot reach out beyond those of the theory's predictions which have been empirically confirmed. Science crucially relies on strategies for understanding how well its theories will fare beyond the limits of current experimental testing. These strategies move beyond empiricism but do not amount to rationalism. They keep faithful to the principle that all knowledge about the world is in the end based on observation. They merely widen the spectrum of observations which are relevant for scientific reasoning and include those observations which are not of a kind predictable by the scientific theories in question. As discussed above, the described strategies of local assessments of scientific underdetermination do not imply scientific realism. However, they do introduce the perspective that arguments based on specific scientific theories could in principle offer epistemic arguments for scientific realism. An understanding of the scientific process that acknowledges the described strategies as an integral part of science thus restores to scientific reasoning the perspective of moving towards a true description of the world while retaining the characterisation of science as an entirely empirics-based enterprise.

The described conclusion may be the most significant and most clear-cut philosophical implication of the new perspective on non-empirical theory assessment. It is not based on specific properties of string theory or other theories but relies solely on the general understanding of the scientific process developed above. The question remains whether a look at specific properties of string theory reveals more far-reaching implications for the realism debate. This will be the topic of the final sections.

7.4 Consistent structure realism

Two properties of string theory are of specific relevance for the question of scientific realism. On the one hand, the fact that string theory generates a final theory claim suggests that a realist perspective may be more natural than in other scientific contexts. On the other hand, the phenomenon of string duality turns out to be adverse to classical forms of scientific realism. In the following, an attempt will be made to mark four cornerstones of a moderate form of scientific realism that accounts for the peculiar conceptual characteristics of string theory. For reasons which will become clear in the course of the ensuing discussion, this conception shall be called "consistent structure realism."

Finality as a perspective on the true theory

Final theory claims offer a far-reaching explanation of predictive success in science. Predictive success can be related to the fact that there is only one universal theory (modulo empirically equivalent variations) that can cover the empirical data available to us today. From there, one can, step by step, move towards the "empirical past" and relax precision and range of the empirical data or reduce the spectrum of physical phenomena to be covered by the theoretical scheme (i.e. reduce the requirements with respect to theory's universality) in order to widen the spectrum of possible theories. Let us assume that the final theory claim holds with respect to empirical data $E(t)$. Predictive success of science then can be expected to occur if, taking steps backwards from $E(t-n)$ to $E(t-n-1)$, the spectrum of possible theories widens only gradually so that there are only a few (bundles of) theories that are compatible with $E(t-n-1)$ but could in principle be empirically distinguishable under $E(t-n)$. Reversing this perspective and moving from the past towards the present, we find scientists approaching the final true theory by decreasing the number of possible alternatives to the true theory allowed at each state of empirical testing. We thus find something like a modified understanding of Peirce's evolution towards trut' that has the final theory as its endpoint. An understanding of scientific reasoning along these lines clearly seems realist in spirit.

Looking at string theory itself as a final theory candidate, we are justified to expect the theory's absolute truth and thereby are entitled to a position that goes beyond the claims of approximate truth associated with a conventional scientific realist perspective. The strength of this realist implication, however, is curbed by the chronically incomplete state of string theory. Due to its incompleteness, string theory as we know it today cannot be understood as one coherent true theory but rather as a multi-leveled web of statements of different status.

Let me specify the current status quo of string physics once more from this perspective. String theory contains mathematical statements which characterize the theory and which are analytically true. Provided that string theory is viable and a final theory, these statements can be applied to the physical world and at that level turn into true statements about the world.

Many interesting claims that are held about string theory today are based on conjectures – mathematical statements for which highly convincing circumstantial evidence is available but no proof can be given. For example, all statements regarding string dualities are of this nature. Since dualities underlie practically all modern research in string theory, it is fair to say that our entire understanding of the theory would collapse if we refused to trust those conjectures. Still, the core conjectures of string theory are considered fairly unassailable. They are supported by a complex web of reasoning. Rigorous proofs often show their viability in certain limits. The accessible characteristics of the overall structure in conjunction with the conjecture's viability in certain limits then hardly seem explicable without assuming the viability of the full conjecture.

The described kind of reasoning bears some resemblance to assessments of scientific underdetermination but does not actually fall into that category. The entire analysis is of a mathematical nature and does not refer to any empirical data. Moreover, it would be misleading to speak of an assessment of underdetermination at all since one asks for an answer to a mathematical question that can only have exactly one correct answer. To give an example, either quantum gravity on anti-de Sitter space is dual to a conformal field theory on its boundary or it is not. The basis on which one decides whether or not to trust the corresponding conjecture is nevertheless in some sense similar to an assessment of underdetermination. One considers all interpretations of the known characteristics of the system which come to mind. Since we only have an incomplete understanding of the system, these could be many, though we know in advance that only one of them can be correct. If only one interpretation has a certain degree of plausibility, one must ask the question whether one believes that the known spectrum of possible interpretations at that level seems complete. If one comes to that conclusion, that is if one believes that one has reasons to take the spectrum of possible interpretations to be limited, one may infer that the only known plausible solution is in fact the correct one. It would be interesting to carry out a thorough analysis of this kind of reasoning, discuss the actual reasoning strategies which provide the basis for having trust in conjectures to the extent it is the case and determine parallels and differences between these strategies and the ones applied in the context of assessment of scientific underdetermination. An investigation of that kind would reach beyond the scope of the present book, however.

In our context, it suffices to point out that the truth of the string theoretical conjectures is not deducible but still considered highly probable. String physical statements based on string theoretical conjectures thus may be added to the body of statements which we can take to be true about the world if the final theory claim is correct.

The understanding of string theory that is based on proved mathematical statements and well-supported conjectures must be expected to remain thoroughly incomplete for a long period of time, however. String theorists only know specific parts and aspects of the theory but often do not fully understand the position and relevance of these parts within the overall scheme. Some characteristics may be misunderstood or incorrectly embedded. Therefore, even if it is taken for granted that the specific statements which characterize certain structural features of the theory are correct, the overall picture presented at this point may be misleading. Correspondingly, our overall picture of string theory is not just incomplete but, even to the extent physicists are able to formulate it, only approximately true at best. The problems connected to establishing the truth of general statements about string theory are indeed reminiscent of the problems related to the status of theories which are not supported by a final theory claim.

Moreover, there are many statements in string physics which are not based on well-established conjectures but amount to well-informed guesswork at this stage. Claims about the number of string theory ground states are one good example for that level of analysis. These claims, which are quite central to our current perspective on the theory, are based on calculations which are done in a certain approximation. It cannot be ruled out, however, that this approximation severely distorts the result. It cannot be excluded either that some new aspect of string theory might eventually correct the present understanding of ground states in string theory and thus render the present calculations irrelevant for the actual ground state selection in string theory. In this light, everyone in the field acknowledges that current calculations just constitute our best guess at the present stage. If future reasoning changes the present understanding, statements like the claim that there are 10^{500} or more string theory ground states could simply be false.

We thus face a wide spectrum of statements on string theory, from rigorously proven purely mathematical statements over well-established conjectures to the application of those statements and conjectures to the world and all the way to kinds of reasoning which merely amount to well-informed guesses. Taking this spectrum of statements into consideration, it would be a clear exaggeration to say that our current understanding of string theory in its entirety is true.

The final theory claim does, however, substantially change the parameters which can be applied in a discussion on the role of truth in science. Endorsing a final theory claim of course implies breaking the pessimistic meta-induction. It enforces the understanding that, when compared to previous states of scientific development, the situation has changed to an extent that not only makes it unlikely that a theory with substantially different ontology will supersede the present one but, far more radically, makes it unlikely that the present theory will be superseded at all.[8]

Moreover, the endorsement of final theory claims on a scientific basis gives an answer to the question as to whether it actually is a goal of science to find true respectively empirically adequate theories about the world. As long as one could hold the perspective that scientific progress was constituted by an infinite sequence of ever more powerful and empirically accurate theories, it was possible to deny the former claim. One could argue that the actual aim of science was a maximization of empirical accuracy without considering full truth or empirical adequacy a realistic or even meaningful goal. Once science explicitly endorses a final theory claim, such a position is no longer compatible with scientific reasoning. Even if the final theory claim fails in the end, holding it at some stage amounts to aiming at truth.

A philosophical endorsement of the final theory claim goes beyond that. It implies that the scientist, in an important sense, has managed to have her hand on the true theory even though she has only a highly insufficient understanding of that theory. All that is needed to arrive at the true theory would be to find a complete and coherent mathematical formulation that incorporates the principles already spelled out. The physicist thus is, to put it pointedly, only a few tautologies away from truth about the world.

---

[8] Note that the refutation of the pessimistic meta-induction is NOT based on a proof that no successor theory can be found. The argument rather works at the same level as the pessimistic meta-induction itself: according to the pessimistic meta-induction it must be reasonably expected that our present theories will be superseded, even though that cannot be proved now. If a final theory claim applies, this situation gets inverted. The reasonable expectation would be that the present theory is a final theory, even though that could not be proved.

Structural realism

As discussed in Section 7.1, an intuitive core argument for scientific realism is based on the understanding that unobservable objects like cells or electrons share so many characteristics of observable objects that a strict ontological distinction between the former and the latter seems implausible and counterintuitive. This argument has a specific ontological implication. It suggests that we should adhere to the understanding that the microphysical objects conjectured by our best theories exist as ontological objects in an external world. This position is often called ontological scientific realism.[9]

There has been a problem with ontological scientific realism for quite some time: arguments supporting it work fairly well close to the limits of observability but get increasingly questionable once one delves deeper into the fabric of fundamental physics. As discussed in Section 5.1, fundamental physics witnesses the progressing decay of formerly stable foundational intuitions about the character of physical objects. The sequence of detachments from classical intuitions was briefly sketched in Section 5.1 and emerged as a specific form of the marginalization of the phenomena. This development renders ontological scientific realism difficult to hold in fundamental physics. By upholding the notion of an external ontological object even after having abandoned our most crucial intuitions regarding the external world and microphysical objects, we might pursue the hopeless enterprise of bolstering against the rising tide the last bastions of a collapsing sandcastle that is bound to be flattened under the next waves of scientific developments.

Stephen French (1998) has pointed out a particularly troublesome problem of the demise of our intuitions in the context of quantum mechanics. Analyzing the concept of the individuality of physical objects, which arguably constitutes an important element of our intuitions about ontological objects, he shows that the question whether or not the concept of individuality can be upheld in a quantum physical context depends on the chosen formulation of the theory. While one formulation allows for a concept of particle individuation, another one does not. Now, he argues, if quantum mechanics strictly implied that particles did not have individuality, the ontological realist could react by retreating to an ontology that does not have that property. If, however, the question as to whether or not a crucial characteristic of objecthood like individuality is upheld depends on

---

[9] A variation of this position that emphasizes the immediate "intuitive" interpretation of experiments as opposed to a realist endorsement of the specifics of theory building goes under the name of entity realism (Hacking, 1983). This position explicitly acknowledges that it may be less applicable in scientific contexts where conceptualization is highly detached from observed objects.

the arbitrary choice of one of the theory's possible formulations, the entire program of ontological realism seems mistaken. After all, the ontology should refer to what there is in the external world, and not to what can be extracted from an arbitrarily chosen mathematical formulation. On that basis, a scientific realism that relies on the posit of ontological objects seems seriously at odds with quantum mechanics.

French's argument has been criticized for being dependent on the framework of non-relativistic quantum mechanics and thereby relying on assumptions which are known to be strictly speaking false. Once one takes special relativity into account and moves on to quantum field theory, it seems very difficult to retain individuality in any formulation of the theory (Malament, 1996; Clifton and Halvorson, 2002). A non-individualist ontology thus seems vindicated at that level. In this light, though quantum physics disfavors an ontologically realist perspective by moving away from our intuitive notion of an ontological object, it does not seem to spell a final verdict on ontological scientific realism itself.

Antirealist implications of a specific theory have also been stressed in the context of general relativity. It has been argued by various philosophers that general relativity is difficult to reconcile with spacetime substantivalism, that is the understanding that points of the spacetime manifold have real existence independently from the objects placed in spacetime.[10] Though spacetime sub-stantivalism constitutes the most straightforward form of a realism about space-time, a rejection of spacetime substantivalism does not per se block moderate forms of ontological scientific realism, however. Recent work on a coherent realist interpretation of general relativity (Esfeld and Lam, 2008), though presented under the name "moderate structural realism," may indeed be under-stood as a moderate form of ontological realism that does introduce a spacetime ontology but avoids the pitfalls of a rigid substantivalist perspective.

Even though neither quantum physics nor relativity thus seem to deal a death blow to ontological realism, they do contribute to the general impression that ontological realism runs counter to the overall character of conceptual develop-ments in fundamental physics. On that basis, they have provided one main motivation for the development and increasing popularity of structural realism, which adheres to scientific realism without relying on ontological objects. Structural realism was originally proposed by John Worrall (1989) as an answer to Laudan's pessimistic meta-induction.[11] Worrall argues that, while

---

[10] Specifically, such arguments were based on the so-called hole argument (see Earman, 1986; Earman and Norton, 1987).

[11] The core idea of structural realism, as Worrall himself emphasizes, can be traced back to Henri Poincaré.

the ontologies of successful scientific theories indeed may change, their core structural characteristics remain the same. Therefore, one can save scientific realism by formulating it as a realism of structure. James Ladyman (1998) shifted the focus of structural realism by formulating it as an ontic rather than an epistemic approach. For Worrall, the crucial point was to make the epistemic point that we can know the approximate truth about the world at a structural level while we cannot know the approximate truth regarding the world's ontology. Ladyman's ontic structural realism, to the contrary, focuses on the conceptual problem of imputing ontologies to modern physical theories and denies that ontological realism is an appropriate concept for understanding the kind of reality that is suggested by those theories. French's quantum physical argument plays an important role in this line of reasoning. Ontic structural realists thus go beyond the epistemic question as to whether we can identify the correct ontological objects. They assert either that reality does not contain ontological objects at all or that ontological objects only play a secondary role to structure. Various versions of structural realism have been presented and widely discussed in recent years (see e.g. Chakravartty, 2004; Stachel, 2006; Esfeld and Lam, 2008) and represent one of the most promising approaches to realism today. Still, the position faces a number of difficult questions itself which continue to be debated. Is it true that structures are much more stable than ontologies? (Psillos, 1999.) How exactly, if at all, is it possible to stabilize relations without relata in ontic structural realism? (Chakravartty, 1998; Ladyman and Ross 2007.) Is there a substantial difference between structural realism and empiricism? (van Fraassen, 2008.)

As stated above, the arguments against the compatibility of quantum physics or relativity with ontological realism, though quite suggestive, are not beyond all doubt. String theory introduces a new and arguably more far-reaching argument against ontological scientific realism. In a sense, string theory carries the erosion of the concept of the ontological object further than any other physical theory. It has been described in Chapter 1 that duality relations constitute a defining element of string theory. A close look at dualities reveals that they are thoroughly incompatible with ontological scientific realism.[12] As described in Chapter 1, a duality relates two theories which are built on different conceptual foundations and establishes that these theories are empirically equivalent. It is possible to formulate a "dictionary" that translates individual properties and objects of one theory into the substantially different properties and objects of the

---

[12] This point was first made in Dawid (2007) and emphasized recently by Rickles (2011) and Matsubara (2013).

dual theory.[13] An example that played an important role in Chapter 6 in the context of final theory claims is T-duality, which relates one theory to an empirically equivalent theory with an inverted compact dimension. T-duality identifies a string wrapped around a compact dimension in one theory with a string that is not wrapped but moves freely along the compact dimension of inverted size in the other. Furthermore, T-duality identifies theories with elementary objects of different dimensions. As described in Chapter 1, a consistent formulation of string theory must involve D-branes, objects of a higher dimension than the one-dimensional strings. It turns out that D-branes of even dimension in one theory correspond to D-branes with uneven dimension in its T-dual theory. T-duality thus implies that the same phenomenology can be described in terms of theories which differ in size, momentum, topological position and even the dimensionality of their objects.

Another duality relation connects specific types of string theory[14] to an eleven-dimensional theory, where the radius of the eleventh dimension corresponds to the strength of the string coupling in its dual theory. The eleven-dimensional theory, called M-theory, is not a proper string theory (recall that string theories have just ten spacetime dimensions) and not well understood. Its existence, however, can be inferred based on duality arguments.

A very important duality relation is the AdS/CFT correspondence which, as mentioned in Chapter 1, relates a string theory on an anti-de Sitter space[15] to a supersymmetric gauge theory on a space with one less dimension. The question in how far this duality can be generalized to other spacetime geometries continues to be a matter of investigation. AdS/CFT duality implies that the same phenomenology can be described by theories which have different numbers of spacetime dimensions and posit entirely different kinds of interaction. (There is gravity in the anti-de Sitter space but no gravity in the dual theory.)

---

[13] Note that duality relations in string theory are of a fundamentally different character than the wave-particle duality encountered in quantum mechanics. The latter expresses the fact that a certain ontological picture (the dichotomy between objects which behave like waves and objects which behave like classical particles) is inadequate for a characterization of quantum mechanics. Classical wave theory would be an empirically different theory than deterministic point mechanics and neither of the two suffices for characterizing quantum phenomenology. We need aspects of both classical concepts in order to give a full description of quantum objects. In quantum mechanics, duality thus denotes a situation where neither of the two dual (ontological) pictures fully applies and both can only be deployed as approximations. In string theory, to the contrary, the dual theories are empirically equivalent and empirically fully viable.

[14] Type IIB and IIA string theory.

[15] Anti-de Sitter space is a space with constant negative curvature. Cosmologically, it corresponds to the space of an empty universe with negative cosmological constant. Since cosmological data indicates a small positive cosmological constant, there is reason to believe that we do not live in an anti-de Sitter type universe.

The theories connected by the described duality relations posit objects of different topology, momentum and dimension which move in spaces of different dimension, obey different symmetry relations and interact based on different kinds of interaction. Dualities thus connect theories whose elementary objects differ in next to all of their core characteristics. Nevertheless, dualities assert that the theories which differ in so many respects are empirically equivalent. It is a classical defensive move of the ontological realist faced with a theory that is at variance with a classical ontology to reduce the core characteristics of the ontological object in a way that jettisons those characteristics which can no longer be specified in a consistent way within the new theoretical framework. In the face of quantum mechanics, the ontological realist could sacrifice determinism, the exact specification of all observable quantities and maybe individuality in order to defend an accordingly reduced concept of an ontological object along these lines. Faced with duality, this strategy seems hopeless. It would amount to positing ontological objects whose defining characteristics specify neither the objects' dimensions nor their topological position, neither their velocities nor the dimension of the space they move within or the interactions they are subjected to. The scientific realist would be left with a notion of ontological object that were virtually empty and, at any rate, would have retained no resemblance whatsoever with the concept of ontological object the realist had set out to defend.

The only strategy left to the ontological realist would be to emphasize the distinction between epistemic access and ontological truth. The realist would then have to assert that, even though several ontologies are compatible with the same empirical data, only one of those ontologies is true. That is, while both of two dual theories are empirically adequate, only one of them is true and the other one is false.

In specific contexts, the ontological realist could actually bolster this position by circumstantial empirical evidence. In many cases, one of two dual theories looks more natural and better calculable than the other. In cases where S-duality relates a theory with a strong string coupling to another one with a weak string coupling, the weakly coupled theory seems the more tractable alternative. With respect to T-duality, one can make a similar argument. A string theory with a compact dimension very much smaller than the string length has a nearly continuous spectrum of Kaluza–Klein modes (representing the winding numbers of a closed string wrapped around the small compact dimension). It would seem far more natural to understand such a scenario in terms of the T-dual picture where the compact dimension is large and the Kaluza–Klein modes are replaced by the spectrum of allowed momenta of the string along the compact dimension. In cases like these, it may seem plausible to say that there is a natural ontology (which the ontological realist might want to understand as the true ontology) and a rather contrived one.

However, the strategy of selecting one true ontology among those connected by duality relations can be argued against at two levels. First, the intuitive point that one ontology can often be found to be more natural than the other is of doubtful significance. In significant parts of parameter space none of the theories stands out as the one that should be naturally chosen. Moreover, sticking to our natural intuitions may be a treacherous guide that does not lead us towards the most interesting perspective. It may happen that one theory seems more natural in a certain regime based on conventional ways of thinking while the other one provides a clearer picture of some deeper aspects of the mechanisms at work. Such suggestions have indeed been made in the context of string theory and cosmology.[16] The question which of the two dual pictures is the more "natural" one thus may often be a matter of perspective and therefore less obvious than it seems at first glance.

The more fundamental objection against selecting one true ontology has to do with the overall structure of string theory. The fact that different string theories with different elementary ontologies are empirically equivalent constitutes an example of underdetermination by all possible data of the form that was attributed to Quine and van Fraassen in Chapter 2. However, it constitutes an unusual example in one respect. The empirically equivalent (that is, dual) theories are embedded within one overall theoretical conception. A good understanding of all individual theories as well as the relations which connect them is taken by string physicists to be a crucial precondition for understanding the theory at all. The abundance and crucial role of duality relations thus constitutes a highly significant characteristic of string theory. The web of dual "theories" defines string theory in its entirety and it would make no sense to reduce the theory to one specific spatiotemporal "theory." It may be speculated that a more fundamental formulation will be found some day that does not have a spatiotemporal framework at all. In this light, the notion of a spectrum of empirically adequate theories, one of which may be true, does not give an adequate characterization of the situation in string physics. The abundance of duality relations does not reflect a separation between true and empirically adequate ontologies. It rather suggests that the distinction between truth and empirical

---

[16] A recent example is Erik Verlinde's much discussed suggestion of a new understanding of gravitation as an emergent force whose deeper explanation is a principle of minimized entropy (Verlinde, 2011). This suggestion may be understood within the context of AdS/CFT in the sense that, while a gravitational force is the most immediate conclusion from the phenomena we observe, an interpretation which is inspired by the dual picture of the AdS/CFT correspondence that does not deploy a gravitational force is the more enlightening one in some crucial respects.

adequacy is out of place in string physics. Under the assumption that string theory itself is an empirically adequate theory, we face a number of empirically adequate spacetime formulations, none of which could be called false. The natural conclusion thus is to conflate truth and empirical adequacy in the given context. This of course directly contradicts ontological scientific realism. String theory demonstrates even more clearly than quantum mechanics that ontological scientific realism is at variance with the developments of fundamental physics.

The anti-ontological implications of string dualities fit well into the line of reasoning followed by ontic structural realists. Even more clearly than in the case of quantum mechanics, the only chance the realist has is to base her realism on structural aspects of the theory rather than on its ontology of objects. It must be emphasized, though, that string dualities would be incompatible also with a conservative interpretation of structural realism that posits the existence of one real structure in space and time. Since dual theories imply different spacetimes, dualities make it impossible to identify one true structure in true spacetime. In order to reconcile structural realism with string dualities, the entire web of dual theories must be understood as one overall theoretical structure, which enforces an abstract notion of structure that does not rely on a spatiotemporal framework.

The conflation of truth and empirical adequacy

Most forms of scientific realism as well as modern forms of empiricism rely on a distinction between truth and empirical adequacy. Ontological scientific realism relies on singling out a true ontology of objects whose dynamics causes the phenomena we can observe. This construction implies that there might, in principle, be another construal of ontological objects which cause the very same phenomena. If so, even though both constructions would be empirically adequate, only one could be true. Equally, van Fraassen's literary understanding of scientific statements implies a distinction between truth and empirical adequacy. His constructive empiricism distinguishes itself from scientific realism by claiming that science aims at empirical adequacy rather than truth.

String dualities, as already suggested above, indicate a full conflation of the two concepts of empirical adequacy and truth. Dualities provide "translation manuals" of one theory into another and thereby constitute an integral part of the overall theoretical scheme. In a string-theoretical context, empirical equivalence is generally understood in terms of dualities, which means that knowing the duality structure of the theory corresponds to knowing the spectrum of empirically equivalent spatiotemporal representations of the theory. As mentioned above, that kind of knowledge is essential for understanding the theory's structure at a deeper level. If we aim at knowing the full truth about the physical world, we therefore must aim at knowing the duality structure and the spatio-temporal representations which are connected by dualities. If string theory is true, knowing the truth about the world thus amounts to knowing the full scope of theoretical structures which are empirically adequate. The distinction between truth and empirical adequacy therefore loses all significance.

One may have one significant worry about the universal validity of a con-flation of truth and empirical adequacy. As string theory is a quantum theory, its foundations raise the same interpretational questions that have haunted founda-tional quantum physicists and philosophers of physics since the advent of quantum mechanics. Thus, the question arises whether the foundational debate in quantum physics could in the end reintroduce into string physics the need for a distinction between truth and empirical adequacy at the level of interpretations of quantum physics. A discussion of that question must be preliminary at best since the fate of foundational questions at the level of string physics remains entirely unclear at this point. The inclusion of gravity may require a substantial rethinking of the foundations of quantum physics. Since the associated con-ceptual changes are impossible to predict at this point, all that can be done is to assess whether the presently available interpretations of quantum mechanics would introduce the need for a distinction between truth and empirical adequacy if their core ideas could be upheld within a string theoretical framework.

In answering this question, it must first be emphasized that most leading interpretations of quantum mechanics, while of course bound to be consistent with the currently available data in quantum physics, are not strictly empirically equivalent with each other. For that reason, two of the main interpretations of quantum mechanics pose no threat for the conflation of truth and empirical adequacy. Spontaneous collapse models (Ghirardi, Rimini and Weber, 1985) assume that a wave function collapse takes place from time to time due to a corresponding term in the equations of motion. This form of collapse and the resulting reduction of quantum entanglement must be observable in principle, which makes the theory an empirically distinguishable alternative to canonical quantum mechanics. The same is true for Everettian quantum mechanics (Everett, 1957; Barrett, 2011), which suggests that no collapse of the wave function occurs at all and our observable world can be identified with one individual branch of the full wave function. The fact that we observe a macro-scopically classical world is explained by decoherence, which largely decouples the individual branches from each other. Strictly speaking, however, the laws of quantum mechanics imply a small chance of recoherence of branches, which in turn implies small differences to the phenomenology of canonical quantum mechanics. Just like in the case of spontaneous collapse, it is therefore essential to the Everettian approach that it is not empirically equivalent to canonical quantum mechanics.

The third main alternative interpretation of quantum mechanics is the Bohmian hidden parameter model (Bohm, 1952), which posits an empirically inaccessible deterministic substructure underneath the observable stochastic quantum processes. While small statistical deviations from canonical quantum statistics may be expected also in a Bohmian context due to the statistical choice of initial conditions, the Bohmian framework may be framed as an empirically equivalent alternative to canonical quantum mechanics without running into immediate inconsistencies. Therefore, we may understand Bohmian quantum mechanics as a kind of approach that might open up a distinction between empirical adequacy and truth.

However, the arguments discussed in the context of string dualities seem to work against the plausibility of such a perspective. Once duality relations have made it impossible to single out one real ontology, the cogency of singling out one out of two or more foundational interpretations disappears as well. In the absence of a specified true ontology, it makes little sense to attribute truth to some extended structural elements attached to that ontology. The Bohmian alternative, if consistent, would appear as another empirically adequate alternative representation without any further reasons for being singled out as true. The contingency of quantum decisions in canonical quantum mechanics would be translated into the contingency of the initial conditions of the Bohmian universe in a similar way as certain properties are translated into their dual properties in the string-theoretical context. If it turned out that Bohmian hidden parameters – or any other alternative interpretation of quantum mechanics – constituted an empirically fully equivalent alternative to canonical quantum mechanics (that is, whatever one may decide to call "canonical" in a quantum gravitational context), the existence of such an alternative arguably would have to be understood as an extension of the spectrum of empirically equivalent spacetime representations of fundamental physics rather than as a constellation that would raise the question of truth and falsehood with respect to the alternative theories.

None of the above arguments strictly excludes some residual role of a distinction between truth and empirical adequacy in a string-theoretical context. It seems justified to say, however, that no significance of such a distinction is in sight at this point and that the entire framework of string physics seems adverse to a perspective that emphasizes that distinction. Philosophically, this means that it seems misplaced to build substantial philosophical distinctions regarding the understanding of fundamental physical theories on the distinction between empirical adequacy and truth.

Scientificality conditions as a basis for delimiting realism from phenomenalism

Up to this point, the analysis was focussed on the question of truth. Besides the assertion that well-established current scientific theories are probably (approximately) true, scientific realism is committed to a second crucial claim, however. It must establish a level of reality that can be distinguished from the phenomenal level. Without this step, adherence to the approximate truth of scientific statements would not project a meaningful concept of a real world behind the phenomena.

How can such a non-phenomenal level of reality be established? It may help first to spell out how this is done in the case of ontological scientific realism. The ontological realist asserts that the truth of statements on physical objects is a different matter than the truth of statements about the phenomenal implications of those objects. According to the ontological realist, it could in principle be the case that several different theories which imply different constellations at the level of real objects are empirically equivalent. Still, just one of those constellations would be actually realized and, correspondingly, just one of the theories would be true. The other theories would be false, even though all deduced statements regarding the phenomenology could be confirmed. Talking about the truth of statements at the level of real ontological objects thus clearly is not the same as talking about the truth of the corresponding phenomenal statements.

A separation of levels along these lines cannot be achieved based on limitations to scientific underdetermination. Theory differentiation in assessments of limitations to scientific underdetermination relies entirely on empirical data and therefore does not reach beyond empirical adequacy. This is true even for the most far-reaching form of that kind of assessment, the final theory claim. Moreover, as stated above, string theory leaves no room for a distinction between truth and empirical adequacy any more, which means that we can call any empirically adequate theory true in the given context. Though this property of string theory moves assessments of scientific underdetermination closer towards realism in some sense, it fully blocks the option of a separation of levels of description based on the distinction between truth and empirical adequacy.

Scientific realism thus can only be established in a weaker sense by distinguishing a non-phenomenal level of description by other means than the employment of different attributions of truth. Limitations to scientific underdetermination indeed provide a way of doing that. Assessments of limitations to underdetermination rely upon the understanding that there exists a theory that solves the scientific problem in question and satisfies a core of scientificality conditions. Global assessments of underdetermination rely on the stronger assumption that there exists an empirically adequate theory that satisfies a core of scientificality conditions. In order to be capable of predictions, science must choose a level of description where underdetermination is limited. The value of scientific reasoning as a method for making successful predictions thus depends on the fact that scientific statements are placed at a "scientific" level of description where limitations to underdetermination apply. Descriptions at a purely phenomenal level are by definition ad hoc in establishing each observed event individually and therefore provide no basis for limitations to underdetermination. The scientificality conditions do provide such basis and thereby open up a level of description where indications of a lack of alternatives can constitute a reason for believing in a theory. The new level of description thus can be meaningfully separated from the phenomenal level. The difference between the two levels is not established based on differing truth values of corresponding statements but rather on the number of descriptions or theories which can be formulated at each level. The scientific level enforces a scarcity of possible theories and therefore raises the probability that an existing empirically informed theory at that level is true. The choice of a scientific level of description must be motivated by a "minimal realist assumption" that is not concerned with the theoretical description itself but with the scientific framework in general: it must be assumed that there is an empirically adequate description of all observable phenomena that satisfies the scientificality conditions.

Once limitations to scientific underdetermination become sufficiently powerful, the scientific structures we find can play the role of sources for predictions in a similar way real objects are supposed to work in canonical scientific realism. Ontological realism explained the fact that our observations adhere to the prediction of our theory by positing that there existed a level of actual microscopic objects which had precisely the causal implications we observe. The approach of non-empirical theory assessment now explains the same fact by positing that the level of scientific description only allows a small number of structures which are empirically distinguishable. When limitations to underdetermination transcend the local level and develop into final theory claims, the statement becomes more radical: there is just one possible scientific structure left that is compatible with the available data. That structure then is concluded to provide a true description of the world.

While the described basis for a distinction between the two levels is obviously weaker than the one deployed in ontological realism, we can nevertheless observe a number of parallels between the role played by the unique consistent scientific structure in our context and the role played by real ontological objects in ontological scientific realism. Both consistent structure and ontological objects are placed at a level of description that is distinguished from the phenomenal level; both are used for explaining the successful predictions in science; and in both cases a uniqueness claim is deployed for countering an instrumentalist interpretation of science. In the case of ontological realism, the claim that there exists a unique true ontology turns science into an enterprise looking for more than for fitting the data: science is taken to search for the ontological reality that enforces the phenomena we observe. In the case of consistent structure, the presumed uniqueness of consistent structure elevates the search for it to a level that also transcends mere data fitting: the instrumentalist principle implying that the way phenomena happen to turn out determines the theoretical tools for describing them is replaced by the opposite picture that, based on some "initial" data set, phenomena must turn out in a specific way because it is the only way allowed for by a consistent scientific description at all. It seems plausible in this light to understand the posit of a unique consistent structure as a weak form of scientific realism. We want to call that position consistent structure realism.

Consistent structure realism is a kind of ontic structural realism. It is a modest form of the latter since it understands structures in their most general non-spatiotemporal form. It fully subscribes to the structural realist claim that a meaningful identification of real ontological objects is not possible in the context of modern physical theories. It adds the specific emphasis that a realist perspective on structure can only be meaningful in contexts where scientificality conditions strongly limit the spectrum of possible scientific structures. All forms of structural realism face the crucial question as to how they can be distinguished from an empiricist position. Consistent structure realism gives its own answer to that question: the reality of structure emerges from the lack of alternative scientific theories compatible with the data. Establishing a lack of scientific alternatives by applying the strategies of non-empirical theory assessment thus adds the bit of information that reaches out beyond empiricism and thereby implements consistent structure realism.

7.5 Whither physical theory?

String theory and related theories in contemporary high energy physics and cosmology have substantially changed our perspective on the physical structure of the world. The present book aimed at pointing out that these changes are accompanied by similarly significant shifts in our understanding of the role that can be played by a physical theory. The concepts of theory confirmation and theory dynamics have to be understood in very different terms in the context of string theory and related theories than in more traditional physics. The philosophical implications of these shifts are twofold.

First, an analysis of string theory can tell us something about science in general. The rising importance of assessments of scientific underdetermination carries a message for the whole philosophy of science. Assessments of scientific underdetermination have always played a crucial role within the scientific process but were largely neglected by the philosophy of science. The case study of string physics now guides us towards acknowledging their relevance within the scientific process and therefore leads towards a more accurate and coherent understanding of science. Science cannot be fully grasped in terms of the coherence between physical theory and empirical data. Rather, it has to be seen as an enterprise that binds together observations and theoretical assessments at different levels of analysis in order to obtain its results.

Second, string theory itself changes the epistemic position of physical theory. No prior theory has offered similarly strong foundations for non-empirical theory confirmation; and no prior theory has offered a coherent basis for the formulation of final theory claims. These developments contribute to a widening of the gap between theory assessment in fundamental physics and in other fields of science. We have no reason to expect that the power of non-empirical theory assessment in string theory can be transferred to other scientific fields which operate under entirely different epistemic conditions. It is even less plausible to expect final theory claims to emerge in other fields of science. The most far-reaching epistemic implications of string physics should be seen as developments which characterize and distinguish a very particular part of scientific reasoning.

Finally, it must be emphasized once again that the current situation in fundamental physics is too volatile for allowing stable judgements on the actual range and power of the strategies discussed. Future scientific developments may be expected to lead to a more robust understanding of these matters. The main message to be extracted from the physical status quo at this point thus cannot be that we have a fully fledged and fully reliable method of non-empirical theory confirmation or that string theory is in fact a final theory. Rather, the novel message is that strategies of non-empirical theory confirmation and strategies of establishing final theory claims have become part of scientific reasoning. Conclusions which can be reached by following these strategies correspondingly turn into scientifically legitimate claims which can fail or succeed as such. This, on its own, is a rather far-reaching conclusion. If the named strategies indeed stabilize their scientific success, they have the capacity of providing an altogether new understanding of the way science relates to the world.

References

Achinstein, P. (2010) "What to do if you want to defend a theory you can't prove: a method of 'physical speculation,'" *Journal of Philosophy* **107**(1), 35–56.

Antoniadis, I., N. Arkani-Hamed, S. Dimopoulos and G. R. Dvali (1998) "New dimensions at a millimeter to a Fermi and superstrings at a TeV," hep-ph/9804398, *Physics Letters* **B436**, 257.

Ashtekar, A. (1986) "New variables for classical and quantum gravity," *Physical Review Letters* **57**(18), 2244–2247.

Barnes, E. C. (2008) *The Paradox of Predictivism*. Cambridge: Cambridge University Press.

Barrett, J. A. (2011) "Everett's pure wave mechanics and the notion of worlds," *European Journal for the Philosophy of Science* **1**(2), 277–302.

Becker, K., M. Becker and J. H. Schwarz (2006) *String Theory and M-Theory: A Modern Introduction*. Cambridge: Cambridge University Press.

Bekenstein, J. D. (1973) "Black holes and entropy," *Physical Review* **D7**(8), 2333–2346.

Bern, Z., L. J. Dixon and R. Roiban (2007) "Is N=8 supergravity finite?," *Physics Letters* **B644**, 265–271, hep-th/0611086.

Bird, A. (2007) "Inference to the only explanation," *Philosophy and Phenomenological Research* **74**, 424–432.

Bohm, D. (1952) "A suggested interpretation of the quantum theory in terms of 'hidden variables,'" *Physical Review* **85**, 166–179.

Bovens, L. and S. Hartmann (2003) *Bayesian Epistemology*. Oxford: Oxford University Press.

Boyd, R. (1984) "The current status of scientific realism," in J. Leplin (ed.) *Scientific Realism*. Berkeley, CA: University of California Press.

Boyd, R. (1990): "Realism, approximate truth and philosophical method," in C. Wade Savage (ed.) *Scientific Theories*, Minnesota Studies in the Philosophy of Science, vol. 14. Minneapolis, MN: University of Minnesota Press.

Brading, K. and E. Castellani (eds.) (2003) *Symmetries in Physics: Philosophical Reflections*. Cambridge: Cambridge University Press.

Butterfield, J. and C. Isham (2001) "Spacetime and the philosophical challenge of quantum gravity," in C. Callender and N. Huggett (eds.) *Physics Meets Philosophy at the Planck Scale*. Cambridge: Cambridge University Press.

191

Callender, C. and N. Huggett (eds.) (2001) *Physics Meets Philosophy at the Planck Scale*. Cambridge: Cambridge University Press.

Capelli, A., E. Castellani, F. Colomo and P. Di Veccia (eds.) (2012) *The Birth of String Theory*. Cambridge: Cambridge University Press.

Carlip, S. (2008) "Black hole entropy and the problem of universality," arXiv:0807.4192.

Chalmers, D. (1996) *The Conscious Mind: In Search of a Fundamental Theory*. Oxford: Oxford University Press.

Chakravartty, A. (1998) "Semirealism," *Studies in the History and Philosophy of Modern Science* **29**, 391–408.

Chakravartty, A. (2004) "Structuralism as a form of scientific realism," *International Studies in Philosophy of Science* **18**, 151–171.

Clifton, R. and H. Halvorson (2002) "No place for particles in relativistic quantum theories?," *Philosophy of Science* **69**, 1–28.

Dawid, R. (2006) "Underdetermination and theory succession from the perspective of string theory," *Philosophy of Science* **73**(3), 298–322.

Dawid, R. (2007) "Scientific realism in the age of string theory," *Physics and Philosophy* **11**, 1–32.

Dawid, R. (2009) "On the conflicting assessments of the current status of string theory," *Philosophy of Science* **76**(5), 984–996.

Dawid, R. (2010) "High energy physics and the marginalization of the phenomena," *Manuscrito* **33**(1), special issue *Issues in the Philosophy of Physics*, 165–206.

Dawid, R. (2013) "String theory and theory assessment," *Foundations of Physics* **43**(1), 81–100.

Dawid, R. (in press) "Novel confirmation and the underdetermination of scientific theory building," in press.

Dawid, R., S. Hartmann and J. Sprenger (in press) "The no alternatives argument," in press.

Douglas, M. (2003) "The statistics of string/M theory vacua," hep-th/0303194, *JHEP* **0305**, 046.

Duhem, P. (1954) *The Aim and Structure of Physical Theory*. Princeton, NJ: Princeton University Press.

Dyson, F. (2008) "The scientist as a rebel," *New York Review of Books*.

Earman, J. (1986) "Why space is not a substance (at least not to first degree)," *Pacific Philosophical Quarterly* **67**, 225–244.

Earman, J. and J. D. Norton (1987) "What price spacetime substantivalism," *British Journal for the Philosophy of Science* **38**, 515–525.

Esfeld, M. and V. Lam (2008) "Moderate structural realism about space and time," *Synthese* **160**, 27–46.

Everett, H. (1957) "Relative state formulation of quantum mechanics," *Review of Modern Physics* **29**, 454–462.

Feynman, R. P. (1950) "Mathematical formulation of the quantum theory of electromagnetic interaction," *Physical Review* **80**, 440–457.

Feynman, R. P. (1969) "Very high energy collisions of hadrons," *Physical Review Letters* **23**, 1415–1417.

Fine, A. (1986) "The natural ontological attitude," in *The Shaky Game*. Chicago, IL: Chicago University Press, pp. 118–119.

Freedman, D. Z., S. Ferrara and P. van Nieuwenhuizen (1976) "Progress toward a theory of supergravity," *Physical Review* **D13**, 3214–3218.

French, S. (1998) "On the withering away of physical objects," in E. Castellani (ed.) *Interpreting Bodies: Classical and Quantum Objects in Modern Physics*. Princeton, MA: Princeton University Press, pp. 93–113.

Galison, P. (1987) *How Experiments End*. Chicago, IL: University of Chicago Press.

Galison, P. (1997) *Image & Logic: A Material Culture of Microphysics*. Chicago, IL: University of Chicago Press.

Georgi, H. and S. L. Glashow (1974) "Unity of all natural forces," *Physical Review Letters* **32**, 438–441.

Gell-Mann, M. (1964) "A schematic model of baryons and mesons," *Physics Letters* **8**, 214–215.

Gervais, J. L. and B. Sakita (1971) "Field theory interpretation of supergauges in dual models," *Nuclear Physics* **B34**, 632.

Ghirardi, G. C., A. Rimini and T. Weber (1985) "A model for a unified quantum description of macroscopic and microscopic systems," in L. Accardi *et al.* (eds.) *Quantum Probability and Applications*. Berlin: Springer.

Glashow, S. L. (1961) "Partial symmetries of weak interactions," *Nuclear Physics* **22**, 579–588.

Goldstone, J. (1961) "Field theories with superconductor solutions," *Nuovo Cimento* **19**, 154–164.

Green, M. B. and J. H. Schwarz (1984) "Anomaly cancellation in supersymmetric D=10 gauge theory and superstring theory," *Physics Letters* **149B**, 117.

Green, M. B., J. H. Schwarz and E. Witten (1987) *Superstring Theory*, 2 vols. Cambridge: Cambridge University Press.

Greene, B. (1999) *The Elegant Universe: Superstrings, Hidden Dimensions, and the Quest for the Ultimate Theory*. London: Jonathan Cape.

Gross, D. J. and F. Wilczek (1973) "Asymptotically free gauge theories 1," *Physical Review* **D8**, 3633–3652.

Guth, A. (1981) "The inflationary universe: a possible solution to the horizon and flatness problems," *Physical Review* **D23**, 347–356.

Hacking, I. (1983) *Representing and Intervening*. Cambridge: Cambridge University Press.

Han, M. Y. and Y. Nambu (1965) "Three triplet model with double SU(3) symmetry," *Physical Review* **139**, B1006–B1010.

Hawking, S. and L. Mlodinov (2010) *Grand Design*. London: Bantam Press.

Hedrich, R. (2007a) "The internal and external problems of string theory," *Journal for General Philosophy of Science* **38**(1), 261–278.

Hedrich, R. (2007b) *Von der Physik zur Metaphysik*. Ontos.

Higgs, P. (1964) "Broken symmetries, massless particles and gauge fields," *Physics Letters* **12**, 132–133; "Broken symmetries and the masses of gauge bosons," *Physical Review Letters* **13**, 508–509.

Horava, P. and E. Witten (1996) "Heterotic & type I string dynamics from eleven dimensions," hep-th/9510209, *Nuclear Physics* **B460**, 506.

Horava, P. (2009) "Quantum gravity at a Lifshitz point," *Physics Review* **D79**, 084008; arXiv:0901.3775.

Howson, C. and P. Urbach (2006) *Scientific Reasoning: the Bayesian Approach*, 3rd edition. La Salle: Open Court.

Hoyningen-Huene, P. (1993) *Reconstructing Scientific Revolutions: Thomas S. Kuhn's Philosophy of Science*. Chicago, IL: Chicago University Press.

Ibanez, L. E. and A. M. Uranga (2012) *String Theory and Particle Physics: An Introduction to String Phenomenology*. Cambridge: Cambridge University Press.

Johansson, L. G. and K. Matsubara (2011) "String theory and general methodology," *Studies in History and Philosophy of Science* **B42**(3), 199–210.

Kahn, J. A., S. E. Landsberg and A. C. Stockman (1992) "On novel confirmation," *British Journal for the Philosophy of Science* **43**, 503–516.

Kaku, M. (1997) *Beyond Einstein: Superstrings and the Quest for the Final Theory*. Oxford: Oxford University Press.

Kobayashi, M. and T. Maskawa (1973) "CP violation in the renormalizable theory of weak interaction," *Progress in Theoretical Physics* **49**, 652–657.

Kovtun, P., D. T. Son and A. O. Starinets (2004) "Viscosity in strongly interacting quantum field theories from black hole physics," *Physical Review Letters* **94**(2005), 111601, hep-th/0405231.

Kuhn, T. S. (1962) *The Structure of Scientific Revolutions*. Chicago, IL: University of Chicago Press.

Ladyman, J. (1998) "What is structural realism?," *Studies in History and Philosophy of Science* **29**, 409.

Ladyman, J. and D. Ross (2007) *Every Thing Must Go*. New York, NY: Oxford University Press.

Lakatos, I. (1970) "Falsification and the methodology of scientific research programs," in I. Lakatos and A. Musgrave (eds.) *Criticism and the Growth of Knowledge*, Cambridge: Cambridge University Press.

Laudan, L. (1977) *Progress and Its Problems*. Berkeley, CA: University of California Press.

Laudan, L. (1981) "A confutation of convergent realism," *Philosophy of Science*, **48**, 19.

Laudan, L. (1996) *Beyond Positivism and Relativism*. Boulder, CO: Westview.

Laudan, L. and J. Leplin (1991) "Empirical equivalence and underdetermination," *Journal of Philosophy* **88**, 449.

Linde, A. (1982) "A new inflationary universe scenario: a possible solution of the horizon, flatness, homogeneity, isotropy and primordial monopole problems," *Physics Letters* **B108**, 389–393.

Lipton, P. (2004) *Inference to the Best Explanation*. London: Routledge.

Maher, P (1988) "Prediction, accommodation and the logic of discovery," *PSA*, **1**, 273–285.

Malament, D. (1996) "In defense of dogma – why there cannot be a relativistic quantum mechanical theory of (localizable) particles", in R. Clifton (ed.) *Perspectives of Quantum Reality*. Amsterdam: Kluwer.

Matsubara, K. (2013) "Realism, underdetermination and string theory dualities," *Synthese* **190**(3), 471–489.

Mattingly, J. (2005) "Is quantum gravity necessary?," in J. Eisenstaedt and A. Kox (eds.) *The Universe of General Relativity: Einstein Studies*, vol. 11. Boston, MA: Birkhäuser.

Murugan, J., A. Weltman and G. F. Ellis (2012) *Foundations of Space and Time: Reflections on Quantum Gravity*. Cambridge: Cambridge University Press.

Musgrave, A. (1985) "Realism versus constructive empiricism," in P. M. Churchland and C. A. Hooker (eds.) *Images of Science*. Chicago, IL: University of Chicago Press.

Nambu, Y. and G. Jona-Lasinio (1961) "Dynamical model of elementary particles based on an analogy with superconductivity 1," *Physical Review* **122**, 345–358.

Norton, J. D. (1993) "The determination of theory by evidence: the case for quantum discontinuity," *Synthese* **97**, 1.

Norton, J. D. (1994) "Science and certainty," *Synthese* **99**, 3.

Pais, A. (1986) *Inward Bound: Of Matter and Forces in the Physical World*. Oxford: Oxford University Press.

Penrose, R. (2005) *The Road to Reality*. London: Vintage Books.

Perlmutter, S. *et al.* (1999) "Measurements of omega and lambda from 42 high-redshift supernovae," *The Astrophysical Journal* **517**(2), 565–586.

Peskin, M. E. and D. V. Schroeder (1995) *An Introduction to Quantum Field Theory*. Reading: Perseus Books.

Pickering, A. (1984) *Constructing Quarks*. Chicago, IL: University of Chicago Press.

Polchinski, J. (1998) *String Theory*, 2 vols. Cambridge: Cambridge University Press.

Polchinski, J. (1999) "Quantum gravity at the Planck length," hep-th/9812104, *International Journal of Modern Physics* **A14**, 2633.

Polchinski, J. (2007) "All strung out?," *American Scientist* **95**(1), 1.

Policastro, G., D. T. Son and A. O. Starinets (2001) "The shear viscosity of strong coupled N=4 supersymmetric Yang Mills plasma," *Physics Review Letters* **87**(2001), 081601, hep-th/0104066.

Politzer, H. D. (1973) "Reliable perturbative results for strong interactions?," *Physical Review Letters* **30**, 1346–1349.

Psillos, S. (1999) *Scientific Realism – How Science Tracks Truth*. London: Routledge.

Putnam, H. (1975) "What is mathematical truth," in *Mathematics, Matter and Method, Philosophical Papers*, Vol. 1. Cambridge: Cambridge University Press.

Quine, W. V. (1970) "On the reasons for indeterminacy of translation," *The Journal of Philosophy* **67**, 179.

Quine, W. V. (1975) "On empirically equivalent systems of the world" *Erkenntnis* **9**, 313–328.

Randall, L and R. Sundrum (1999) "A large mass hierarchy from a small extra dimension" and "An alternative to compactification," *Physical Review Letters* **83**, 3370–3373 and 4690–4693.

Rescher, N. (2000) "The price of an ultimate theory," *Philosophia Naturalis* **37**, 1–20.

Rickles, D. (2011) "A philosopher looks at string dualities," *Studies in the History and Philosophy of Modern Physics* **42**(1), 54–67.

Riess, A. *et al.* (1998) "Observational evidence from supernovae for an accelerating universe and a cosmological constant," *The Astronomical Journal* **116**(3), 1009–1038.

Riordan, M. (1987) *The Hunting of the Quark – A True Story of Modern Physics*. New York, NY: Touchstone.

Roush, S. (2006) *Tracking Truth*. Oxford: Oxford University Press.

Rovelli, C. (1998) "Loop quantum gravity," gr.qc/9710008, *Living Reviews in Relativity* **1**, 1.

Rovelli, C. and L. Smolin (1990) "Loop space representation of quantum general relativity," *Nuclear Physics B* **331**(1), 80.

Salam, A. and J.C. Ward (1964) "Electromagnetic and weak interactions," *Physics Letters* **13**, 168–171.

Scherk, J. and J.H. Schwarz (1974) "Dual models for nonhadrons," *Nuclear Physics* **B81**, 118.

Schweber, S. S. (1994) *QED and the Men Who Made It: Dyson, Feynman, Schwinger and Tomonaga*. Princeton, NJ: Princeton University Press.

Schwinger, J. (1951) "On gauge invariance and vacuum polarization," *Physical Review Letters* **82**, 664–679.

Shimony, A. (1970) "Scientific inference," in R. C. Colodny (ed.) *Pittsburgh Studies in Philosophy of Science*, Vol. 4. Pittsburgh, PA: University of Pittsburgh Press, pp. 79–172.

Sklar, L. (1975) "Methodological conservatism," *Philosophical Review* **84**, 384.

Sklar, L. (1981) "Do unborn hypotheses have rights?," *Pacific Philosophical Quarterly* **62**, 17–29.

Sklar, L. (2000) *Theory and Truth*. Oxford: Oxford University Press.

Smolin, L. (2006) *The Trouble with Physics*. Boston, MA: Houghton Mifflin.

Stachel, J. (2006) "Structure, individuality and quantum gravity," in D. Rickles, S. French and J. Saatsi (eds.) *The Structural Foundations of Quantum Gravity*. Oxford: Oxford University Press, pp. 53–82.

Stanford, P. K. (2001) "Refusing the devil's bargain: what kind of underdetermination should we take seriously?," *Philosophy of Science* **68** (Proceedings), 1.

Stanford, P. K. (2006) *Exceeding our Grasp – Science, History, and the Problem of Unconceived Alternatives*. Oxford: Oxford University Press.

Stanford, P. K. (2009) "Scientific realism, the atomic theory and the catch-all hypothesis: can we test fundamental theories against all serious alternatives?," *British Journal for the Philosophy of Science* **60**(2), 253–269.

Strominger, A. and C. Vafa (1996) "Microscopic origin of the Bekenstein–Hawking entropy," hep-th/9601029, *Physics Letters* **B379**, 99.

Strominger, A. (1998) "Black hole entropy from near horizon microstates," *JHEP* **9802**, 009, hep-th/9712251.

Stöltzner, M. (1999) "Vienna indeterminism: Mach, Boltzmann, Exner," *Synthese* **119**, 85–111.

Susskind, L. (2003) "The anthropic landscape of string theory," hep-th/0302219.

Susskind, L. (2006) *The Cosmic Landscape*. New York: Little Brown.

't Hooft, G. (1971) "Renormalizable Lagrangians for massive Yang–Mills fields," *Nuclear Physics* **B35**, 167–188.

't Hooft, G. and M. J. G. Veltman (1972) "Regularization and renormalization of gauge fields," *Nuclear Physics* **B44**, 189–213.

van Fraassen, B. C. (1980) *The Scientific Image*. Oxford: Clarendon Press.

van Fraassen, B. C. (2002) *The Scientific Stance*. New Haven, CT & London: Yale University Press.

van Fraassen, B. C. (2008) *Scientific Representation: Paradoxes of Perspective*. Oxford: Oxford University Press

Veneziano, G. (1968) "Construction of a crossing-symmetric, regge behaved amplitude for linearly rising trajectories," *Nuovo Cimento* **57A**, 190.

Verlinde, E. (2011) "On the origin of gravity and the laws of Newton," *JHEP* **1104**, 029.

Weinberg, S. (1967) "A model of leptons," *Physical Review Letters* **19**, 1264–1266.

Weinberg, S. (1992) *Dreams of a Final Theory.* London: Panthcon.

Weinberg, S. (1996) *The Quantum Theory of Fields.* Cambridge: Cambridge University Press.

Weinberg, S. (2001) *Facing Up.* Cambridge, MA: Harvard University Press.

Weingard, R. (1989) "A philosopher's look at string theory," reprinted in Callender, C. and Huggett, N. (2001) *Physics Meets Philosophy at the Planck Scale.* Cambridge: Cambridge University Press.

Wess, J and B. Zumino (1974) "Supergauge transformations in four dimensions," *Nuclear Physics* **B70**, 39–50.

Witten, E. (1995) "String theory dynamics in various dimensions," hep-th/9503124, *Nuclear Physics* **B443**, 85.

Witten, E. (1996) "Reflections on the fate of spacetime," reprinted in C. Callender and N. Huggett (eds.) (2001) *Physics Meets Philosophy at the Planck Scale.* Cambridge: Cambridge University Press.

Woit, P. (2006) *Not Even Wrong: The Failure of String Theory and the Continuing Challenge to Unify the Laws of Physics.* London: Jonathan Cape.

Worrall, J. (1989) "Structural realism: the best of both worlds?," *Dialectica*, **43**(1–2), 99.

Worrall, J. (2000) "The scope, limits, and distinctiveness of the method of 'deduction from the phenomena': some lessons from Newton's 'Demonstrations' in optics," *The British Journal for the Philosophy of Science* **51**, 45.

Wüthrich, C. (2004) "To quantize or not to quantize: fact and folklore in quantum gravity," *Proceedings PSA 2004: Contributed Papers.*

Yang, C. N. and R. L. Mills (1954) "Conservation of isotopic spin and isotopic gauge invariance," *Physical Review* **96**, 191–195.

Zwiebach, B. (2004) *A First Course on String Theory.* Cambridge: Cambridge University Press.

Index

Printed in the United States
By Bookmasters